The
Mathematics
Bible

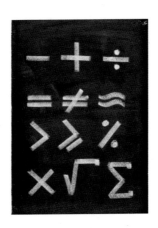

The
Mathematics
Bible

The Definitive Guide to the
Last 4,000 Years of Theories

Colin Beveridge, PhD

FIREFLY BOOKS

A FIREFLY BOOK

Published by Firefly Books Ltd. 2016

First printing

Publisher Cataloging-in-Publication Data (U.S.)

A CIP record for this title is available from the Library of Congress

Library and Archives Canada Cataloguing in Publication

A CIP record for this title is available from Library and Archives Canada

Published in the United States by
Firefly Books (U.S.) Inc.
P.O. Box 1338, Ellicott Station
Buffalo, New York 14205

Published in Canada by
Firefly Books Ltd.
50 Staples Avenue, Unit 1
Richmond Hill, Ontario L4B 0A7

Interior design Malcolm Parchment

Printed in China

Conceived, designed, and produced by
by Cassell, a division of Octopus
Publishing Group Ltd
Carmelite House, 50 Victoria
Embankment, London EC4Y 0DZ
www.octopusbooks.co.uk
Editorial Director Trevor Davies
Editor Pollyanna Poulter
Art Director Yasia Williams-Leedham
Production Controller Meskerem Berhane
Colin Beveridge asserts his moral right to
be identified as the author of this work.

CONTENTS

INTRODUCTION

"God made the integers; all else is the work of man."

Leopold Kronecker

"When a mathematical or philosophical author writes with a misty profundity, he is talking nonsense."

Alfred North Whitehead

Mathematics isn't *just* about knowing your times tables and logarithmic laws. Its history is full of stories and characters, legends and parables — and I've done my best to bring them to life.

The course of the history of math is as convoluted and intricate as a complicated historical novel: heroes being exiled (I lost count of the number of mathematicians who

Sir Isaac Newton, scientist, alchemist mathematician and feuder.

left Europe in the 1930s), ferocious feuds (there's a real humdinger between Newton and Liebniz), atrocious skulduggery (did Évariste Galois meet his end as the result of a deceitful plot?), and moments of revelation (one of which led Sir William Rowan Hamilton to vandalize the nearest bridge).

Just like a novel borrows ideas and themes from oral history and older

Évariste Galois met his doom in a duel as a result of atrocious skulduggery.

texts, *The Mathematics Bible* owes a great deal to the many people who have taught me, shared their favorite stories, and pointed me at puzzles, games and paradoxes. I'm especially grateful to:

Enigma Code Machine from World War II.

machine works, and for checking my work;

- Josie Damen-Lane, for showing me that quaternions had actual proper uses;

- Henriette Finsterbusch, for being the person I imagine I'm talking to if I can't explain something;

- Dave Gale, my *Wrong, But Useful* podcast co-host, for convincingly pretending to like statistics so that we can squabble about them;

- Adam Goucher, for pointing me at his elliptic curve nomogram;

- Samuel Hansen, podcast genius and Abel Prize nominee, for smoothing the path for opinionated and unreliable histories of math with his series *Relatively Prime* (relprime.com);

- T.K. Briggs, Officer for Long Curly Hair (and Education) at Bletchley Park for showing me how an Enigma

- Sally Maltby, for getting me to think about the Poincaré disk;

- Chris Maslanka, whose puzzles were (and remain) a key part of my mathematical education;

- Barney Maunder-Taylor for telling me to tell stories so that this isn't a terrible book. I've tried;

- John O'Connor and Ed Robertson, who got me interested in the history of mathematics and will presumably shake their heads at my many errors and misconceptions, as they always did with my essays;

- Matt Parker and Colin Wright for starting MathJam and giving me the chance to meet *amazing* mathematicians;

- Christian Perfect-Lawson, Katie Steckles and Peter Rowlett, for starting *The Aperiodical* and giving me the chance to write about amazing mathematicians;

- Brian Rodrigues and Phil Stonehouse, teachers whose main lesson was that the tangents were more interesting than the curves;

- Hugo Rowland and the other students who patiently listened to my excited rambling about Difference Engines and approximating π;

- Martin Stellar, for taking me to the Alhambra to see the Escher exhibition. I can't wait to go back!

Alhambra, Granada, Spain, well worth a visit or two!

I consider myself lucky to have an extremely supportive family:

I owe a debt of thanks to my uncle Bill Beveridge, for telling me the Lorenz story at an impressionable age, and telling me to read *Gödel, Escher, Bach;* I plan to do the same one day for my sons, Bill Beveridge Russ and Frederick Andrew Russ. I'm grateful to my partner, Laura, whose descent into the Wiki-vortex prompted the question "Have you ever heard of the Scottish Café?" I hadn't. (Oh, and for her unflagging support and encouragement. That also helped.) Her mother, Nicky, looked after said young Bill for much of the time I was writing this, turning it from an impossible project to just a difficult one.

A goose is integral to the story of the mathematicians who frequented the Scottish Café.

I'm also thankful, as always, to my parents (Linda Hendren and Ken Beveridge) and my brother (Stuart Beveridge), for cheering me on and mocking me whenever I grumble.

The White Rabbit from Alice in Wonderland, *probably the most popular math book of all time.*

CHAPTER 1
THE DEADLY TRIANGLE AND HOW MATH DEVELOPED

When you learn about math, it's natural to wonder, "Who came up with this stuff?"
The answer, usually, is, "Someone anonymous, a long, long time ago."

This chapter tries to provide a little more enlightenment with a bit of the "where" and
the "when." There were some mathematical characters back then, though, and I'll
introduce you to a few of them.

Many of the equations and diagrams on this
whiteboard make sense; however, there are some
deliberate errors. See how many you can find!

MATH BEFORE WRITING

In the beginning, there was the null set. After that, things became confused.

There's an obvious difficulty in trying to describe math before writing: we don't know very much about it, because it wasn't written down.

It's easy to speculate about how numbers came about: a hunter returns to his village to tell of the animals he's been tracking. Is it worth sending a party after them? You need to know how many there are. How big are they? What kind of animal?

This isn't a purely human thing: every so often, you'll read in the more lurid papers that "scientists say" horses

Some scientists believe that horses can count.

can do math. Scientists are given a special deference by the media; "scientists say" being used in a way that "politicians say" or "sportsmen say" never are.

The point is not simply that a horse can demonstrate counting skills by clomping a hoof on the ground a certain number of times — having a sense of quantity is a vital survival skill.

From counting, it's a shortish step to keeping tally, not perhaps in the way that an accountant would but as an essential part of everyday life for early humans. How many cattle went up to the mountain, and how many came back? How many days until the rains come?

The first artefact that seems likely to be a mathematical object is the 20,000-year-old Ishango bone, discovered in 1960 close to the Semliki River near what is now the border between Congo and

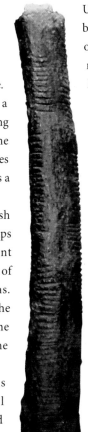

The 20,000-year-old Ishango bone is thought to have been used as a calculator.

Uganda. At first it was thought to be a "tally stick" — the equivalent of an accountant's ledger — but it may actually have been used as a kind of calculator.

The bone is from the leg of a baboon, and has a piece of quartz at one end. It is covered in markings, which appear to be numerical in nature. In the central column there is 3, 6, 4 and 8, along with what may be 10 and 5 — a series of doublings and halvings. In the left column, the numbers 11, 13, 17 and 19, all of the prime numbers between 10 and 20; in the right column are marks that may represent 9 (10–1), 19 (20–1), 21 (20+1) and 11 (10+1).

The first unambiguously mathematical objects are from Sumeria (now Iraq), around 3,000 BCE — the same time as Stonehenge, which it's hard not to see as a mathematical object in its own right.

Wedge-shaped cuneiform writing used by the Babylonians.

BABYLONIAN MATH

Some of the earliest mathematical writing that survives comes from the Babylonian empire, around 1900 BCE.

The Babylonians, who overran the Sumerians in what is now Iraq, carved wedge-shaped symbols into clay, which were then baked solid in the sunshine, giving them permanent records of transactions as well as a kind of table for frequently used calculations. One of the things they carved into such tablets were tables of the square numbers. If you know the square numbers, you can multiply fairly easily without doing anything more complicated than subtracting and dividing by 4 — if you want to multiply two numbers together, you can square their sum, square their difference, find the difference between those squares and divide it by 4.

For example, to work out 7×18, the Babylonians might find:

$(7 + 18)^2 = 625$ and $(18 - 7)^2 = 121$

The difference is 504, and a quarter of that is 126 (which, you might want to check, is indeed 7×18).

Similarly, tables of reciprocals were used to carry out division, and reference tables for solving certain cubic equations have been found.

There is also good evidence that the Babylonians had a strong understanding of Pythagoras's Theorem, long before Pythagoras did.

The Babylonian Tablet Plimpton 322 from between 1900BCE and 1600BCE may display an understanding of Pythagoras's Theorem.

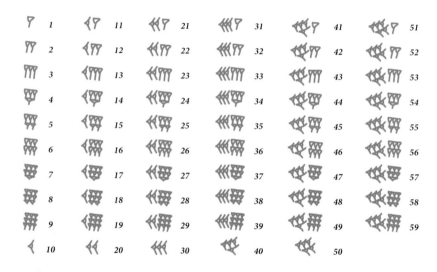

The "wine glass" and "eye" symbols that make up the Babylonian numerals 1–59.

BASE 60

The arithmetic you and I learned in school and use for almost everything every day is base 10 — when you get to nine, you run out of digits and start a new column.

In base 10, every digit in the number has a value 10 times more or less than its neighbors.

However, it wasn't always that way — some of the first written systems, used by the Sumerians about 3000 BCE and the Babylonians from about 1830 BCE, used 60 as a base, not 10. You can still see hints of this in the way we measure time and angles — there are 360 degrees in a circle, which is 6 × 60, and each degree can be divided into 60 minutes, each of 60 seconds, just like with time.

The Babylonian system didn't have 60 separate symbols for each possible digit, but broke things down into groups of 10 on the way.

From one to nine, you'd draw the shape for the number 1, ending up with what looked like a stack of wine glasses. Ten looks a bit like a cartoon eye looking to the right. A number like 47 would have four eyes, followed closely by seven stacked wine glasses; 63 would

be a single wine glass (representing 60) followed by three wine glasses representing the three.

The Babylonians did have the idea of a zero, but only between numbers — for example, 7,247 (which is $2 \times 60^2 + 47$) would be written as two wine glasses (for two), then a big gap (representing no 60s) and then the symbol for 47 (as explained above).

Confusingly, the system didn't differentiate between 1, 60 and 3,600, all of which would be represented as a single glass! You'd have needed to understand from the context what was meant.

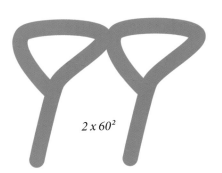

2×60^2

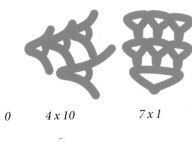

0 $\quad$ 4×10 $\quad$ 7×1

Babylonian numerals showing the number 7,247.

EGYPTIAN MISCONCEPTIONS

It's often asserted that Egyptian builders needed precise right-angles to construct pyramids, and to measure these, they used pieces of rope 12 units long.

They would arrange them into a triangle like the one pictured. Even in those days, the good old 3-4-5 was a famous example of a right-angled triangle.

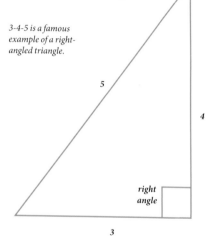

3-4-5 is a famous example of a right-angled triangle.

There are, however, problems with that. There's one medium-sized archaeological problem: no such rope triangles have ever been found. There's also a big, practical problem: I don't know if you've ever tried to use a rope- or string-based set square, but they're really not very effective. They're stretchy. They're inaccurate. They're completely impractical for small-scale building projects, let alone enormous ones like pyramids.

It's not clear where, precisely, the builders of the pyramids got their right-angles from — but it almost certainly wasn't from 3-4-5 triangles.

EGYPTIAN MATH

The ancient Egyptian system of writing down numbers — especially fractions — looks cumbersome, but has a certain internal logic.

Whole numbers were somewhat similar to Roman numerals: to write 245, you'd write down the symbol for

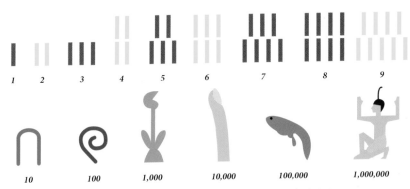

The Egyptian system for writing down numbers was cumbersome but logical.

100 twice, the symbol for 10 four times, and the symbol for 1 five times.

2 x 100   *4 x 10*   *5 x 1*

Where we might write an approximation to π as 3.1415 — which is to say, $3 + \frac{1}{10} + \frac{4}{100} + \frac{1}{1,000} + \frac{5}{10,000}$ — the Egyptians would most likely prefer $3 + \frac{1}{8} + \frac{1}{60}$; apart from $\frac{2}{3}$ and sometimes $\frac{3}{4}$, the only fractions allowed were ones with 1 on top. These were denoted by an eye sitting on top of the number.

In comparison with Greek mathematics, Egyptian math was extremely practical. Whereas Greek math focussed on proof for the sake of proof and complete abstraction, the Egyptian problems that have survived show us that they tended to examine things like dividing bread between workers or finding the area of a tract of land.

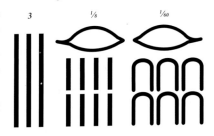

3   *⅛*   *¹⁄₆₀*

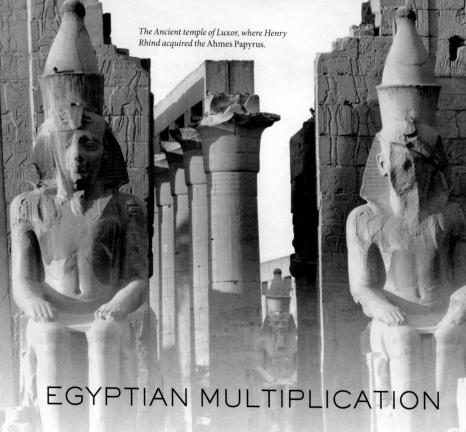

The Ancient temple of Luxor, where Henry Rhind acquired the Ahmes Papyrus.

EGYPTIAN MULTIPLICATION

The Ahmes Papyrus, *possibly the oldest surviving math worksheet, was bought by Henry Rind in Luxor in 1858 — it's sometimes also known as the* Rhind Papyrus.

Ahmes was the scribe; the scroll dates from about 1650 BCE, although Ahmes claims to be copying a document some 200 years older.

The document displays a lovely method of doing long multiplication — for example, to work out 61 × 85, you would repeatedly double 85 until the left-hand column is over half of 61, like this.

1:	**85**
2:	**170**
4:	**340**
8:	**680**
16:	**1,360**
32:	**2,720**

You'd then work out which numbers in the left column added up to 61:

$$32 + 16 + 8 + 4 + 1 = 61$$

Adding up the corresponding numbers in the right-hand column …

$$2,720 + 1,360 + 680 + 340 + 85$$

… gives your answer, which is 5,185 — the correct answer. Dividing the numbers works in a similar way — and both methods make use of the fact that it's easy to add and subtract Egyptian numerals, but nothing more complex.

The Ahmes papyrus (which is six meters long, so there's plenty of room on it) also has a table showing how to double fractions: you might think that the best way to write $2 \times \frac{1}{7}$ would be $\frac{1}{7} + \frac{1}{7}$, but the Egyptians didn't like that: instead, they'd write it as $\frac{1}{4} + \frac{1}{28}$.

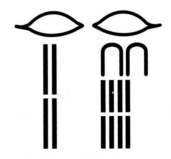

Total of sum, in the Egyptian system of writing down numbers.

There is some logic there: the first fraction is a decent approximation to $\frac{2}{7}$ in its own right (about 14 percent off, in modern terms), while $\frac{1}{7}$ is nowhere near. This lookup table made using the multiplication method straightforward, even with fractions!

THE TRIANGLE OF DEATH

The story goes that, in ancient Greece, one of the members of the Pythagorean School had come up with something odd.

The Pythagoreans believed that every number could be written as a fraction, but Hippasus had been looking at the square root of 2.

Although he didn't have modern algebraic notation available to him, the gist of his argument was this: if the square root of 2 is a fraction, you must be able to write it, in simplest form, as …

$$a/b$$

… and if you can do that, then …

$$a^2/b^2 = 2$$

… or, if you like …

$$a^2 = 2b^2$$

… which means a has to be even. That's fine, but it means there's a number c such that $a = 2c$, and you can write …

$$a^2 / b^2 = 4 c^2 / b^2 = 2$$

That simplifies to $2c^2 = b^2$, so b has to be even as well.

That's a problem: if a and b are both even, then a/b isn't in its simplest form — which led Hippasus to conclude that the assumption that you can write the square root of 2 as a fraction was wrong.

Pythagoras, it is said, looked carefully at the proof, asked a few questions to

The Pythagoreans, seen here celebrating the sunrise, believed in balance and harmony.

verify it was true, and unceremoniously threw Hippasus into the Mediterranean to drown. Not even Pythagoras liked a smarty pants.

The Pythagoreans, you see, believed in balance and harmony, especially in mathematics, and introducing doubt — the notion that there were numbers that they could not explain — was a dangerous idea that could have undermined the sect.

PYTHAGORAS

There's not a lot known for sure about Pythagoras of Samos (c. 570 BCE to c. 495 BCE). Most accounts of his life were written many years after his death.

Born on the Greek island of Samos, when he was about 40 Pythagoras moved to Kroton, now Crotone in modern-day Italy, to start a sect.

What the sect did is clouded in secrecy, but it's generally agreed that they studied math, music and astronomy, while maintaining a strict regimen of abstinence.

At some point, the populace of Kroton rose up against the secretive (and hugely influential) society and burned down their meeting-place.

One thing that is certain: Pythagoras wasn't the first person to discover or even prove Pythagoras's Theorem. It was well-known to the Babylonians, who used it in a way that suggests they

Pythagoras established his sect in what is now Crotone, Italy.

had a proof, even if the proof itself hadn't yet been found. It suited Plato to attribute it to Pythagoras, so he did.

Musically, he's credited with Pythagorean tuning, in which notes are tuned in the ratio of 3:2 (a perfect fifth

PITAGORA

Bust of Pythagoras, philosopher, mathematician and scientist.

apart), and it's said that he believed there were nine planets — however, this is probably because he liked the number 10 and counted the Sun among the number.

It's not clear how or when Pythagoras died (myths abound). The town of Pythagoreion on Samos is named in his honor.

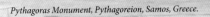
Pythagoras Monument, Pythagoreion, Samos, Greece.

PYTHAGORAS'S THEOREM

In a right-angled triangle, the square on the hypotenuse is equal to the sum of the squares on the other two sides.

It's almost certainly the most famous theorem in mathematics. It's usually shortened, these days, to $a^2 + b^2 = c^2$, even if that's rather imprecise.

It was almost certainly used and understood to some extent long before Pythagoras — Babylonian, Mesopotamian, Chinese and Indian mathematicians all made sense of it independently.

Although it's not clear that Pythagoras was the first to prove the theorem, he did come up with an especially nice one, pictured below.

Arranging four copies of the original triangle makes a pair of squares — the outer one has a side length of $(a+b)$ and the inner one, c. Moving the diagonally-opposite triangles together leaves you with two squares, one with side length

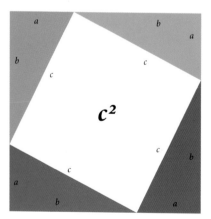

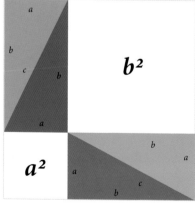

Former teacher James Abram Garfield became the 20th President of the United States.

of a and another with side length b. Since the triangles and the outer square haven't changed size, the area of the original smaller square (c^2) and the total area of the two final squares ($a^2 + b^2$) must be the same.

There are many, many proofs of Pythagoras's Theorem — one book contains 370: some geometric, some algebraic, some involving calculus.

One was contributed by James Garfield, who went on to become President of the United States.

About binomial theorem I'm teeming with a lot o' news, With many cheerful facts about the square of the hypotenuse.

Gilbert & Sullivan

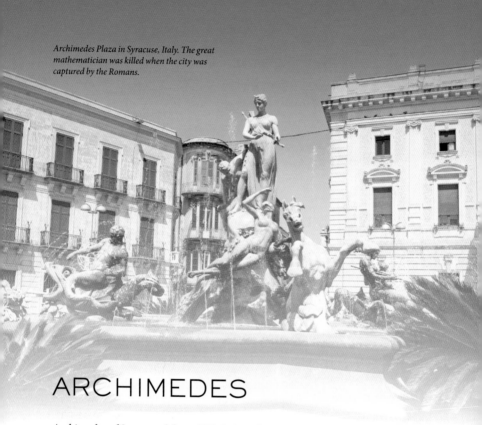

Archimedes Plaza in Syracuse, Italy. The great mathematician was killed when the city was captured by the Romans.

ARCHIMEDES

Archimedes of Syracuse (about 287–212 BCE) is generally held to be one of history's most important scientists.

In his *Men of Mathematics*, E.T. Bell puts Archimedes alongside Newton and Gauss at the head of the mathematical field.

There are some great legends attached to Archimedes, although it's not clear whether they all actually happened — they've been embellished

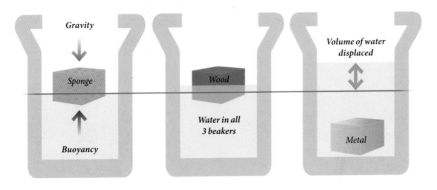

Archimedes realized that he could calculate the volume of an object by measuring the amount of water it displaced.

with every retelling over the last two and a half millennia.

The most famous one is the bath. Everyone has a mental image of Archimedes running naked through the streets of Syracuse screaming "Eureka! Eureka!" after realizing you could determine the volume of an object by dunking it in water and measuring the overflow.

This allowed King Hiero II to determine that a dishonest goldsmith

had cheated him in making his crown.

The second Archimedean legend is that he organized the people of Syracuse with polished shields to reflect the Sun's heat onto the Roman fleet, setting it on fire.

Recent experiments have failed to replicate the Archimedean heat ray, though — and it's been pointed out that catapults or arrows would have been a far more efficient way to set a fleet alight.

Telling a Roman soldier to be careful of his circles may have been the last words spoken by Archimedes.

Another (and more plausible) piece of weaponry he's reputed to have invented is The Claw — a sort of balance with (as the name suggests) a claw attached to one end. Dropping the claw end onto a ship and letting gravity take its course would pull the ship out of the water and/or sink it.

The last legend concerns his death. Despite his best efforts, Syracuse was eventually overrun by the Romans. Although the Roman soldiers had orders not to harm Archimedes, he reportedly told a careless soldier not to step on his circles; the Roman took offence and killed him. An alternative version has him carrying mathematical instruments, which a soldier took for weapons. In any event, he didn't survive.

This barely scratches the surface of Archimedes's achievements — among other things, he also invented a screw for pumping water uphill, supposedly pulled a ship out of the harbor using a system of pulleys, built an orrery (a working model of the solar system), improved the catapult and invented the odometer for measuring distances traveled.

Mathematically, he came up with a method for approximating π, got quite a long way toward discovering

Archimedes, working on one of his war inventions.

calculus, managed to find the area between a parabola and a line using geometric series, and estimated how many grains of sand would fit into the universe.

His face is on the Fields Medal, and his catchphrase — Eureka! — is the state motto of California.

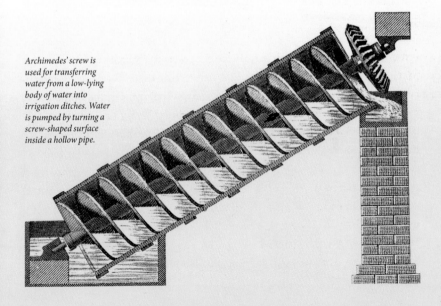

Archimedes' screw is used for transferring water from a low-lying body of water into irrigation ditches. Water is pumped by turning a screw-shaped surface inside a hollow pipe.

EUCLID'S ELEMENTS

*Somewhere around 300 BCE, in Alexandria, Egypt, a mathematician wrote a book —
or rather, 13 books.*

These would remain the de facto mathematics text for well over two millennia. It wasn't until the 1830s that people started to pick holes in his work, and Euclid's *The Elements* would remain in general use into the 20th century.

Practically nothing is known about Euclid, except that he was working at the same time that Ptolemy I was in charge — the scant biographical details were recorded several centuries later.

The Elements isn't a completely original work, but it is the first time everything known about math at the time was written down in one place. It was the *Basic Math For Dummies*

*Statue in honor of Euclid,
Oxford University Museum
of Natural History.*

of its day, setting everything out in a neat, logical order, and showing how each proposition follows from the one before.

Euclid insists on proving everything — except for a small number of ideas he says are self-evident common notions or axioms, things assumed without proof, such as "a line may be drawn through any two points," and "if one thing is the same as a second, and the second thing is the same as the third, then the first and third are the same."

One of these assumptions, the parallel postulate caused an awful lot of trouble. It's much less elegant than the other assumptions,

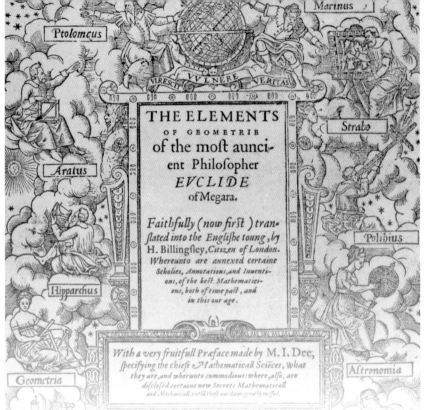

The title page of Sir Henry Billingsley's 1570 English version of Euclid's The Elements.

and many people put a lot of time into trying to prove it based on the others; unfortunately for them, that can't be done.

Although it's best known as a work on geometry, *The Elements* also contains work on number theory — Euclid's proof that there are infinitely many prime numbers is a timeless classic in its field.

In the work, he also examines perfect numbers, factorizing, and an algorithm for finding the highest common factor of two numbers.

BYRNE'S ELEMENTS

Oliver Byrne (c. 1810–1890) was an English mathematician, and surveyor of the Falkland Islands. He's best-known for an astonishing book, The First Six Books of the Elements of Euclid.

Now, obviously, Byrne didn't write *The Elements* — that boat had sailed more than 2,000 years before he was born — but Byrne did revise Euclid so that instead of dull and impenetrable proofs involving angle ABC and line segment OP, it was full of colorful and impenetrable proofs, with the shapes in question being inserted in the text.

It's a gorgeous piece of work, although probably no more practical than any other translation of Euclid. The mathematics blog *The Aperiodical* uses Byrne's depiction of a circle theorem as its logo.

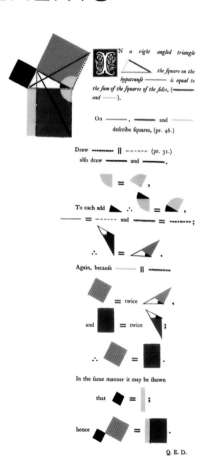

A page from Oliver Byrne's illustrated version of the work of Euclid.

INFINITELY MANY PRIMES

Euclid's proof that there are infinitely many prime numbers — numbers like 2, 5, and 17 that have no factors other than 1 and themselves — is one of the best-known in math. No section on Euclid would be complete without it, so here's my interpretation.

You start (as in many of Euclid's proofs) by assuming the opposite of what you want to prove — if you end up with something absurd, then your assumption must be wrong, and the thing you're trying to prove must be right.

In this case, that means you assume there are finitely many prime numbers, so there's a highest prime number.

Next, multiply together all of the numbers up to the highest prime number to get a Big Number. (Don't actually do the sum, just imagine you've worked out this enormous number.) Then, add one, to get a Bigger Number.

This Bigger Number isn't a multiple of 2 (because it's one more than your Big Number which had 2 as a factor), nor a multiple of 3 (three was also a factor of your Big Number), nor of any other number up to and including your largest prime.

But that's a problem! This Bigger Number must have prime factors, but it definitely doesn't have any factors (apart from 1) in your list of primes. Either the Bigger Number is a prime number, or it has prime factors that are larger than what you assumed was the largest prime.

Either way, your assumption is clearly incorrect, so the opposite is true: there is no biggest prime number, so there must be infinitely many of them.

Every number x Big Number + 1 = Bigger Number

Eastern Arabic	٠	١	٢	٣	٤	٥	٦	٧	٨	٩
Western Arabic/ European	0	1	2	3	4	5	6	7	8	9

Eastern Arabic numerals compared to modern Western Arabic/European numbers.

AL-KHWARIZMI'S COMPENDIOUS BOOK

Muhammad ibn Mūsā al-Khwārizmī (about 780–850 CE) was a Persian scholar at the House of Wisdom in Baghdad under the Abbasid Caliphate.

Al-Khwārizmī's work on Indian numerals (now often called Arabic numerals) was some of the first works on decimals to be translated into Latin, meaning he's an important step in the decimal system's spread across the world.

His book, *The Compendious Book on Calculation by Completion and Balancing* was one of the first to explain precisely how to solve linear and quadratic equations using algebra — which is itself a corruption of the Arabic word al-jabr, part of the book's title, meaning "restoring" or "completion."

The word *algorithm*, incidentally, comes from a corruption of al-Khwārizmī.

Al-Khwārizmī didn't use letters — or even numbers — the way we do now. Instead, everything was written out in words: if you think "$x^2 + 2 = 3x$" is hard to understand, try "a thing squared, plus two, gives thrice the original thing," but the operations of completion (adding the same thing to both sides of an equation) and balancing (subtracting the same thing from both sides) are just about the first things you learn to do in high-school algebra today.

ALHAMBRA PALACE AND ISLAMIC ART

The small fortress of the Alhambra, in Granada, southern Spain, was originally built in 889CE.

Rebuilt in the 11th century, it became a proper royal palace under Sultan Yusuf I about 300 years later. After the conquest of Granada by the Spanish in 1492, the palace fell into ruins again, and only in the last 200 years has it been restored to its glorious, eye-twisting best.

It's an imposing structure, up high on the hill overlooking the city and, architecturally, I'm sure it's extremely interesting. But that's not why it's important to me; it's important to me because of the walls.

View of Alhambra at sunset, Granada, Spain.

Everywhere you look, there's a new geometric pattern, some regular and soothing, others with symmetries you struggle to pin down. Many (possibly all) of the 17 wallpaper groups can be seen, and no two rooms feel remotely alike. M.C. Escher visited the Alhambra in 1922, and it's fair to say he got something out of the visit. The symmetry and tessellation of the palace's tiles are clearly reflected in his later art, and it's quite usual when walking through the rooms of the palace to wind up

Tile decoration, Alhambra, Spain

Columns surrounding the lions fountain in the Alhambra, Granada, Spain.

somewhere it seems impossible to have got to from where you started. A look at some of the arches calls Belvedere to mind — who could say which pillar was behind which?

A UNESCO World Heritage Site, the Alhambra Palace is very high on my list of places to revisit with a better camera. It's a must for anyone who's ever covered a sheet of paper in geometric doodles.

WALLPAPER GROUPS

Any repetitive two-dimensional pattern that covers a plane belongs to one of 17 wallpaper groups — kinds of meta-pattern that describe the symmetries in the design.

There are four kinds of symmetry allowed in a wallpaper group:

- A translation: the whole pattern can be moved.

- A rotation: the whole pattern can be twisted around a point (either by a half-turn, a quarter-turn, a third-turn or a sixth-turn).

- A reflection: the whole pattern can be flipped across a line, as if making a mirror image.

- A glide reflection: a combination of a translation along a line and a reflection in it.

These four symmetries can be combined in many ways, from the simple (the group p1, for instance, is an asymmetric pattern repeated over and over again, in two different directions) to the complicated (like p6mm, which has sixfold rotational symmetry and reflections in six different directions).

Wallpaper patterns can be found all over the world.
A: Egypt. B: Tahiti. C: Assyria. D: China.

If the idea of repetitive, symmetrical tilings leaves you cold, you can always investigate aperiodic tilings, such as the Penrose tilings, named after the mathematician and physicist Roger Penrose (1931–). These display glimpses of symmetry, but have no large-scale symmetry. In a spectacular instance of a mathematical discovery being a solution in search of a problem, aperiodic tilings turned out to be excellent models for quasicrystals, discovered in 1982 by Dan Shechtman — who was awarded the Nobel Prize for his work in 2011.

The Incan khipu system of cords of different lengths with various knots was used to keep records.

KHIPU

It looks a little like a jellyfish — a piece of corded cotton, with several further cords hanging from it. These cords themselves may have cords hanging from them; some *khipu* have as many as 10 or 12 levels of subsidiary cords — when spread out, they can look a little like a mind map.

During the Inca period, *khipus* (*khipu* is the Cusco Quechua word for knot) were used for record-keeping and transporting messages across the Andes (most of the surviving ones date from around the 15th century). The knots in the cords are somewhat

Leadlight in form of a Penrose pattern - a specific geometric figure in mathematics.

A knot-master was required to decipher the khipu, *seen here with a* yupana *(an Incan abacus) on the left.*

mysterious, although some clearly represent a decimal numbering system — a *khipumakayuq*, or knot-master, could be summoned to court to read documentary evidence from *khipus* as proof of payments or ownership. Other knots are just as clearly not numerical, and there's all manner of debate about what they represent — and precisely how (they certainly don't appear to be a transliteration of anything in the Quechua language.)

Most fascinatingly, Gary Urton has decoded what he believes to be a sort of postal code for a village — using a three-dimensional co-ordinate system, recorded on a *khipu* long before Descartes rediscovered the idea.

Sadly, many *khipu* were destroyed by conquistadors, who saw them as anti-Catholic.

CHAPTER 2
THE RENAISSANCE,
THE NEGATIVE
AND THE IMAGINARY

A tide of math sweeps over Europe, bringing with it negative numbers and their imaginary square roots, an astrologer correctly predicts his own death, a monk anticipates Mandelbrot, and two mathematicians in Renaissance Italy duke it out over cubic equations.

The 13th-century monk Udo of Aachen was hailed as one of the greatest mathematicians of all time.

MATH BEFORE WRITING

The story of modern mathematics in Europe starts, realistically, with Leonardo of Pisa (c. 1170–1250), better known as Fibonacci.

Fibonacci's father, Gugielmo Bonacci ("*Fibonacci*" means "son of Bonacci") was a wealthy trader, and it was on his travels with his father that young Leonardo came across Arabic numerals in Algeria.

Statue of Leonardo Fibonacci in Pisa, Italy.

This was a revelation: he realized this would make arithmetic an awful lot easier, and he spent the next 20 years studying under the great Arab mathematicians around the Mediterranean at the time.

When he came back to Italy at the start of the 13th century, he wrote a summary of what he'd learned, the *Liber Abaci*, or *Book of Calculation* — the first European book to champion Arabic numerals. Demonstrating how much easier bookkeeping, conversions and interest calculations were with Arabic numerals, the book revolutionized the study of math in Europe.

Naturally, Fibonacci isn't primarily remembered for that contribution, but for something he didn't invent, the *Fibonacci sequence* (which was known in India some 500 years earlier).

The number of spirals in a sunflower's display of seeds can be a Fibonacci number.

In a fictionalized rabbit population, he asks what happens under certain assumptions — the size of each generation being the sum of the two previous generations, the sequence starts 1, 1, 2, 3, 5, 8, 13 ... it's a lovely sequence — the ratio between successive terms gets progressively closer to the Golden Ratio, φ (about 1.618).

Although it's not as ubiquitous as some people would have you believe, many artists (including Salvador Dalí) have used it in their work, and it does give rise to a rather pleasing spiral.

It's also, in a very loose sense, the "most irrational" number — the hardest to get a good fractional approximation to — which is why, if you count the number of spirals in the seeds of sunflowers or the scales of pineapples, you usually find a Fibonacci number.

He also worked on Diophantine equations, and has an identity named after him, the Brahmagupta-Fibonacci identity: if you multiply two sums of two square numbers together, you get another sum of two squares — for example:

$$(100 + 4)(49 + 81) = 104 \times 130 = 13{,}520 = 2{,}704 + 10{,}816 = 52^2 + 104^2$$

LUCA PACIOLI AND THE SUMMA DE ARITHMETICA

Luca Pacioli (c. 1447–1517) didn't invent much mathematically, although he did apparently invent double-entry accounting and the Rule of 72, a rule of thumb for determining how quickly an investment will double in size.

What he did was write a big textbook summarizing the state of mathematics at the end of the 15th century — for the first time, written in a local dialect rather than in Latin. This made it more widely accessible than most books of the time, and opened up mathematics to a much wider audience.

It's also significant for its notation — it used p. and m. in place of + and –, and R. in place of √, but is otherwise quite mathematically legible. He has a look at quartic equations (summarizing which ones he knows how to solve), and gives a method for approximating square roots. He also analyses several games of chance, anticipating Fermat

Title page of Summa de arithmetica, geometria, proportioni, et proportionalita *by Luca Pacioli, 1523.*

and Pascal, although his analysis is off the mark.

A few years later, Pacioli worked with Leonardo da Vinci on a book called *Da divina proportione*, a study of the

Diuina proportione

Opera a tutti glingegni perspicaci e curiosi necessaria Oue cia studioso di Philosophia prospectiua Pictura Sculptura

Pacioli worked with Leonardo da Vinci on Da divina proportione.

math of proportion and perspective, including the golden ratio. I'm a bit jealous — don't get me wrong, I love the illustrators I'm working with for this book, they're brilliant, but they're not *Leonardo* brilliant!

THE RULE OF 72

If you invest $100 at 2 percent interest, it takes about 36 years for your investment to double in size. If you invested it at 8 percent interest, it would take roughly nine years. In general, if you invest your money at n% interest, it takes about $72/n$ years to double it.

$100 invested at 2% = $200 in 36 years
$100 invested at 8% = $200 in 9 years
x invested at n% = $2x$ in $72/n$ years

Pacioli most likely discovered this rule empirically (to work it out algebraically needs logarithms, which he certainly didn't have access to), but it's a remarkably accurate rule of thumb.

THE IMPROBABLE LIFE OF GIROLAMO CARDANO

When people talk about someone being a Renaissance man or woman, they mean someone with talents across the spectrum, rather than limited to a single field. Girolamo Cardano (1501–1576) could well be the model they're thinking of.

He was a doctor, the first to describe typhoid fever. He was a mathematician, obviously. He was a gambler — a card shark and a chess fiend. He was an astrologer and reputedly predicted the date of his own death, although he just as reputedly committed suicide, which skews the odds slightly. He was a philosopher, and he was the author of well over 200 books, which add music, physics and religion into his list of subjects. He invented the combination lock, the gimbal (which allows a gyroscope to

Italian physician and mathematician, Girolamo Cardono.

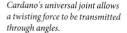

rotate in three dimensions) and the universal joint — and he was one of the first to recognize that deaf people were capable of reading and writing without learning to speak.

His mathematical achievements are a bit special: he was the first to come to grips with negative numbers, and had a decent stab at imaginary numbers too; he wrote about how to solve cubics and quartics, and wrote the first book exploring the rules of probability — although it wasn't published until a century after he wrote it, by which time several others had duplicated his efforts independently.

His life, though, was a complete soap opera: as an illegitimate child (his dad was pals with Leonardo), he struggled to gain admittance to the College of Physicians — although this may have had as much to do with his belligerent attitude.

Cardano's universal joint allows a twisting force to be transmitted through angles.

THE RENAISSANCE, THE NEGATIVE AND THE IMAGINARY

Once he began practicing as a doctor, he traveled all the way to Scotland where he cured the Archbishop of St. Andrews of asthma, which had previously left him unable to speak. Despite the advantages he gained from being the only gambler alive to understand probability (as well as being an expert in cheating), he

Cardano studied at the University of Pavia in Italy.

Cardano was at a huge advantage as a gambler in that he understood probability.

somehow contrived to be perpetually penniless. Things went from bad to worse when his eldest son was executed for poisoning his wife, and from worse to even worse yet when he had to turn in his other son, Aldo, for stealing from him, leading to his son's banishment. Things went from worse yet to worse still when

Cardano was tried for heresy in 1570 after publishing a horoscope of Jesus. He eventually patched things up with the church, and was given an annuity by Pope Gregory XIII.

Cardano died in 1576, and is best remembered for a method of solving cubic equations.

THE FACTORIZATION CHALLENGE

If Renaissance Italians hadn't been so obsessed with keeping their methods secret, the story of how we came to solve cubics (such as $6x^3 - 31x^2 - 7x + 60 = 0$) would be far less interesting.

It was Scipione del Ferro, a math professor at the University of Bologna who made the first known breakthrough some time around 1515, figuring out how to solve equations that look like:

$$x^3 + 5x = 6$$

These are cubics with no square numbers, and no negative numbers anywhere — negative numbers weren't a thing in Europe at that point. If he had known about negative numbers, his method would have been enough to solve any cubic: if you replace the variable x with something else cleverly enough, you can make the squared term disappear, reducing it to the form he knew how to solve.

Del Ferro was good at math and probably better at keeping secrets: he didn't tell anyone that he knew how to solve cubics until just before he died, when he revealed his method to his student Antonio Fior. Fior wasn't especially good at math or at secrets, and, almost immediately, rumors began to circulate that the cubic had been solved.

Sometimes, just knowing that a solution to a problem exists makes the problem a whole lot easier: Nicolo Tartaglia ("the stammerer"), inspired by the rumors, figured out how to solve equations like:

$$x^3 + 5x^2 = 6$$

These equations were without an x term, but Tartaglia could handle

Courtyard of Archiginnasio in the University of Bologna where Scipione del Ferro taught math.

them — and he didn't care who knew it. His method, though? Top secret.

Fior didn't like this development. I imagine him stomping a foot when challenging Tartaglia to a slow, mathematical duel: each would set the other 30 cubic equations to solve over the next couple of months. Fior, having stomped, now rubbed his hands together and cackled: he believed Tartaglia could only solve equations without the x term, so he set him equations of the other type.

Sadly for Fior, Tartaglia had extended his method to cover all cubic equations just a week before the contest started, so he rattled off all of the answers within a couple of hours, winning the competition handily.

Cardano was curious about Tartaglia's secret method, and pleaded and cajoled until Tartaglia finally gave in and revealed it — on condition that Cardano not reveal it until Tartaglia himself did. Cardano, naturally, did the dirty on Tartaglia, somewhat duplicitously arguing that because his student (Lodovico Ferrari) had worked out how to solve quartic equations, of which cubics are a special case, the method was clearly now in the public domain.

There is one wrinkle in Tartaglia's method, which Cardano deals with in

HIERONYMI CAR
DANI, PRÆSTANTISSIMI MATHE
MATICI, PHILOSOPHI, AC MEDICI,
ARTIS MAGNÆ,
SIVE DE REGVLIS ALGEBRAICIS,
Lib.unus. Qui & totius operis de Arithmetica, quod
OPVS PERFECTVM
inscripsit,est in ordine Decimus.

HAbes in hoc libro, studiose Lector, Regulas Algebraicas (Itali, de la Cos-
sa uocant) nouis adinuentionibus,ac demonstrationibus ab Authore ita
locupletatas,ut pro pauculis antea uulgò tritis,iam septuaginta euaserint. Ne-
q́ folùm , ubi unus numerus alteri, aut duo uni,uerùm etiam,ubi duo duobus,
aut tres uni æquales fuerint,nodum explicant.   Hunc aũt librum ideo seor-
sim edere placuit,ut hoc abstrusissimo, & planè inexhausto totius Arithmeti-
cæ thesauro in lucem eruto, & quasi in theatro quodam omnibus ad spectan-
dum exposito, Lectores incitarétur,ut reliquos Operis Perfecti libros, qui per
Tomos edentur,tanto auidiús amplectantur,ac minore fastidio perdiscant.

Cardano, Tartaglia and del
Ferro worked in secret, jealously
guarding their methods.

the traditional way of sweeping it under the carpet and noting that it all comes out okay in the end. In certain circumstances, you have to take the square root of a negative number. Cardano, perhaps alone among his contemporaries, was happy enough with negative numbers, but their square roots? He invited his readers to ignore the mental tortures associated with them.

Artis magnae, sive of algebraicis regulis, *better known as* Ars magna *(Latin for great work), is an important mathematical book originally written in Latin by Gerolamo Cardano in 1545.*

The final insult to Tartaglia? Although Cardano credited both him and Ferrari in *Ars Magna*, where he explained the methods, the formula used to this day for solving cubics is universally known as Cardano's method.

BOMBELLI AND THE IMAGINARY

There are two complementary strands of mathematical development: on one hand, there are pure mathematicians who develop math for the sake of it, entirely indifferent to whether it has a use.

Celebrated mathematician Raphael Bombelli has a crater on the moon named after him.

O n the other, there are engineers and scientists who find that the mathematical tools available aren't adequate, so they develop their own, ignoring the mathematicians shouting "but that's *impossible!*" on the grounds that it works. Rafael Bombelli (1526–1572) sits firmly in the latter camp.

Bombelli was the first to suppose that the nonsensical numbers coming out of Cardano's method for solving cubic equations were mathematical objects you could work with. In 1572, just before he died, he released a book called *L'Algebra*, taking the convoluted explanations of algebra available at the time and turning them into something that could be understood by people who didn't have a top-class education.

Although Cardano was the first to make sense of negative numbers, Bombelli's *L'Algebra* was the first time anyone in Europe had written down the rules for working with them.

Negative arithmetic got off to a poor start, frankly; Bombelli was the first to say the dread phrase "minus times minus is a plus," which still plagues teenagers nearly five centuries later.

$$-6 \times -6 = 36$$

Modern math students would be perplexed by the way Bombelli expressed his rules.

Even though he introduced one of my biggest bugbears, I forgive him entirely, because he went on to discuss complex numbers in a way that's not all that far removed from how I'd describe them today. The only significant difference is that, in modern math, the square root of -1 is denoted by i, rather than Bombelli's "plus of minus." I can hardly imagine what today's teenagers would make of rules expressed like that.

He realized that the square root of a negative number didn't work exactly like a regular number — they were neither positive nor negative, and needed their own rules.

Revisiting methods for solving cubics and quartics like del Ferro's that had been abandoned because of these persistent impossible numbers, Bombelli found he could make them work using the theory of complex numbers he'd developed.

UDO OF AACHEN

Udo of Aachen (c. 1200–1270) was a Benedictine monk, scholar, poet and mathematician. His best-known poetical work is Fortuna Imperatrix Mundi, *usually known by its choral title,* O Fortuna *from* Carmina Burana.

That's not even the most remarkable thing about Udo. When professor Bob Schipke visited Aachen cathedral in 1999, he noticed something incredible about a nativity scene. The Star of Bethlehem appeared to be in the shape of the Mandelbrot set.

He tracked down some of Udo's

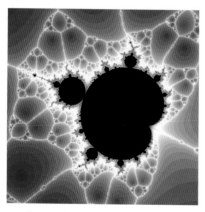

Mandelbrot sets can be used to generate wildly psychedelic computer images.

original writings, which had been discovered in the 19th century and immediately filed away, presumably by a curator with no mathematical training. In them, Udo described the basics of probability theory, performed Buffon's needle experiment, and wrote down rules for working with "profane" and "spiritual" numbers — which correspond precisely to what we call "real" and "imaginary" numbers. The process of repeatedly multiplying and adding them together, as in the Mandelbrot set, was seen as an allegory to determining who would be drawn to God, and who would be cast into darkness.

Udo of Aachen was a remarkable mathematician, many centuries ahead of his time, making use of techniques that weren't widely adopted in Europe for many decades to come …

... Or at least, he *would* have been. Sadly, or perhaps luckily, Udo didn't exist: he was an elaborate April Fool's joke created by writer Ray Girvan in 1999.

Like all of the great hoaxes, it contains just enough detail to be plausible — and just enough "what on earth was I thinking when I believed this nonsense?" when the truth is revealed.

The cathedral in Aachen, where Bob Schipke was supposedly inspired to research the venerable Udo.

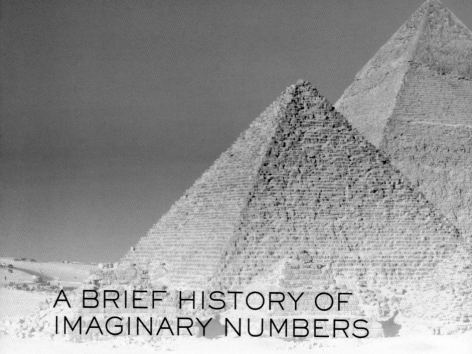

A BRIEF HISTORY OF IMAGINARY NUMBERS

Around 50 CE, Heron of Alexandria, for reasons that are still unclear, tried to find an impossible cross-section of a pyramid.

When it turned out he'd need to work out the square root of 81–144, he sensibly gave up: the idea of negative numbers wasn't one available to him, let alone the idea of finding their square roots. However, he deserves a little bit of credit for coming up with the first known case of an imaginary number.

Even after Bombelli treated them with the respect they deserve, they were still seen as something not quite right. It was Descartes, writing about them the following century, who gave them the name "imaginary": it was an appeal to mathematicians to stop making things up that weren't real!

Heron of Alexandria came up with the first known imaginary number.

Understanding negative numbers would have helped Heron with his pyramid problem.

As with most things, it took Leonard Euler to set things on the right path. He introduces them early in his *Elements of Algebra* and uses them without any apology. Euler was the first to spot that you could write, $e^{i\theta} \equiv \cos(\theta) + i \sin(\theta)$, which gives rise to Euler's identity:

$$e^{i\pi} + 1 = 0$$

This is frequently voted mathematics' most beautiful equation, linking five of the most important constants in math (e, i, π, 1 and 0) as well as the three most important operations (addition, multiplication and exponentiation), not to mention an equals sign, around which the whole edifice is built.

WHAT COMPLEX
NUMBERS ARE FOR

For several centuries, complex numbers were almost wholly of theoretical use — they made it easier to solve cubic and quartic equations, they tidied up the theory behind polynomials (if you include complex numbers, every polynomial of degree n with complex coefficients has n zeros), they simplified the expression of sines and cosines ... all very neat, but hardly necessary.

There are two main places in physics where complex numbers eventually turned out to be unavoidable: in the theory of circuits, and in the murky world of the quantum.

If you work with a direct current (DC) circuit, you can work out the links between potential difference, current and resistance without needing complex numbers — real numbers

Complex numbers are not required when working with direct current.

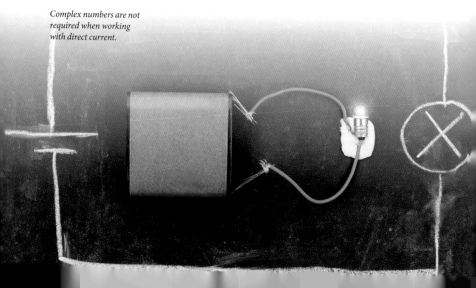

rule the DC world. However, as soon as you introduce an alternating current, you get effects called *inductance* and *capacitance*, which complicate matters.

You *can* write down real-valued equations that link everything together, but they're extremely convoluted. It's much, much simpler to link resistance, inductance and capacitance together as a complex-valued quantity called the *impedance*, giving you a straightforward equation linking the current, potential difference and impedance — which looks an awful lot like the DC version with current, potential difference and resistance.

Where they're *completely* indispensable, though, is in quantum physics. It's almost unthinkable to tackle the idea of probability waves without complex numbers — in *principle*, one could probably come up with a framework to tackle the subject without them, but physicists and mathematicians alike would see it as an unnecessary perversion.

Complex numbers can be used to simplify calculations relating to alternating current.

THE ARGAND DIAGRAM

There's a lovely description of imaginary numbers as being "at right angles to reality," which makes all sorts of sense if you think of the real and imaginary parts as coordinates on a map.

The x-axis corresponds to the real part of the number, and the y-axis to the imaginary part. The *modulus* of a complex number, meanwhile, is just how far the corresponding point is from the origin, (0,0), and its *argument* is the angle between the x-axis and the line from the origin to the point.

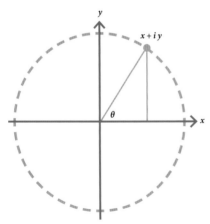

An Argand diagram is a plot of complex numbers as points $z = x + iy$

This kind of map — an Argand diagram — is named after Jean-Robert Argand (1768–1822), although it was first conceived by Caspar Wessel (1745–1818). It's widely used in complex analysis to figure out what's going on, in just the same way as sketches help solve geometry problems. For example, the points of the Mandelbrot set are plotted on an Argand diagram.

Thinking about complex numbers geometrically helps conceptualize them, especially if you're more of a visual thinker: adding or subtracting two complex numbers together is as simple as putting the lines that represent them end-to-end and seeing where the combination ends; multiplication is trickier (it's a combination of adding the angles together and multiplying the lengths), but it can still give an insight.

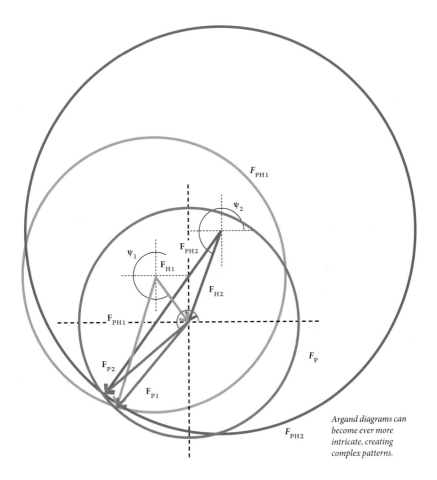

F_{PH1}

ψ_2

F_{PH2}

ψ_1

F_{H1}

F_{H2}

F_{PH1}

F_P

F_{P2}

F_{P1}

F_{PH2}

Argand diagrams can become ever more intricate, creating complex patterns.

Anything you can do with a regular graph you can (within reason) do with an Argand diagram. Equations of lines and circles are most elegant in complex form, and the whole field of contour integration is nicely opened up by these pictures.

CHAPTER 3
ANCIEN RÉGIME

A real monk goes hunting for perfect numbers, a fake knight tries his hand at gambling, and a real lawyer (who pretends he's not a real mathematician) has margins just a little too narrow.

Mathematics can be a gambler's best friend.

DIOPHANTUS OF ALEXANDRIA

To get to grips with math in 17th-century France, we need to start in 3rd-century Greece, with Diophantus of Alexandria (c. 201–285).

To the casual mathematician, he might be recognized as the subject of a riddle:

Arithmetica is Diophantus's most famous work.

Here lies Diophantus, the wonder behold.
Through art algebraic, the stone tells how old:
God gave him his boyhood one-sixth of his life;
One twelfth more as youth while whiskers grew rife;
And then yet one-seventh ere marriage begun;
In five years there came a bouncing new son.
Alas, the dear child of master and sage,
After attaining half the measure of his father's life chill fate took him.
After consoling his fate by the science of numbers for four years,
he ended his life.

To the best of my knowledge, there's no reason to think the riddle is biographical.

I'm more troubled by the poet running out of steam in the last couple of lines and abandoning scansion and rhyme! But it may be the kind of puzzle Diophantus would have appreciated. His most famous work, *Arithmetica*, is a collection of puzzles we'd now call algebraic.

Arithmetica sets and solves many problems, both with known values and unknowns (for this reason, Diophantus is sometimes called the Father of Algebra; I prefer to think of him as a beloved great-uncle), each of which required a specific technique to solve. Although Diophantus didn't use negative numbers or irrational numbers (which he thought absurd), he was one of the first to accept fractions as possible solutions.

Largely forgotten between the end of the days of Classical Greece until the 15th century or so, *Arithmetica* was eventually translated by Bombelli and, most famously, by Bachet. It was this edition that Pierre de Fermat owned.

Today, equations that only involve polynomials and whole numbers are called Diophantine equations after him.

Pompey's Pillar in Alexandria dates from around the time when Diophantus lived in the city.

SOLVING DIOPHANTUS'S RIDDLE

It's not too difficult to work out the ages of Diophantus and Diophantus Junior from the riddle — all you have to do is use a little common sense and add a few fractions.

From the riddle we can work out that the father lived for a sixth of his life, followed by a twelfth of his life, followed by a seventh of his life, followed by five years, then a half of his life (while his son was alive), then another four years.

Adding up the fractions gives:

$$\frac{1}{6} + \frac{1}{12} + \frac{1}{7} + \frac{1}{2} = \frac{14 + 7 + 12 + 42}{84} = \frac{75}{84} = \frac{25}{28}$$

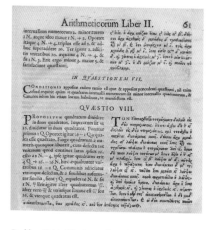

Problem II.8 in Diophantus's Arithmetica *inspired 17th-century French mathematician Pierre de Fermat.*

So, only three twenty-eighths of his life isn't counted in the fractions. We know that is the remaining nine (5 + 4) years of his life. If three twenty-eighths of his life is nine years, one twenty-eighth of that life is three years, so a full life is:

3 x 28 = 84 years.

Diophantus, according to the riddle, died aged 84, so his son lived to 42.

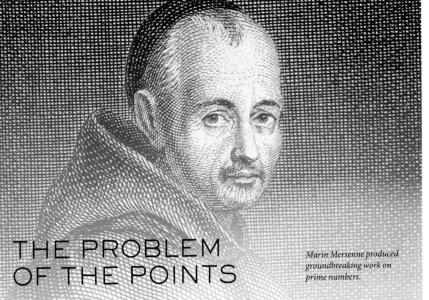

THE PROBLEM OF THE POINTS

Marin Mersenne produced groundbreaking work on prime numbers.

Marin Mersenne (1588–1648) wasn't really a mathematician, but a theologian with a healthy interest in the mechanics of music. The thing is, in the early 17th century, there wasn't really a cultural difference between the three subjects: a person of learning would switch from one to the other with ease.

In 1635, Mersenne established an informal academy for the discussion of theology and the sciences; his Parisian Academy comprised 140 or so correspondents, including René Descartes and Pierre de Fermat.

In his time, he was a celebrated music theorist — he was one of the first to write down the laws governing the vibrations of a fixed string (that its frequency is inversely proportional to its length, to the square root of the tensile force, and inversely proportional to the linear density of the string). He can also claim to have invented the reflective telescope, although he probably didn't realize the significance of what he was doing.

That's not what Mersenne is known for, mathematically at least. He's known for the Mersenne primes, prime numbers of a particular form: all the prime numbers $M_p = 2^p - 1$, where p is a prime number. For example …

$$\mathbf{M_3} = 2^3 - 1 = 7$$

Portal of the former Jesuit College Henri IV in La Flèche, France, where Mersenne studied.

… which is a prime number. Mersenne listed the first few ps that he thought worked: 2, 3, 5, 7, 13, 17, 19, 31, 67, 127 and 257. It's the kind of list that would receive not many marks in an exam: it contains errors, and there's no indication of how it was worked out.

There are five errors: M_{67} (which is of the order of 150 quintillion) is, in fact, a composite number as is $M_{257} \approx 2.3 \times 10^{77}$; meanwhile, he overlooked M_{69}, M_{89} and M_{127}, all of which are prime. That's not to say these were obvious mistakes — the primality of them was only established in the late 19th century, and the correct list wasn't completely certain until well into the 20th.

Mersenne primes are comparatively easy to find — there's a simple test (the Lucas-Lehmer test) to see if there is a factor. This means that the largest primes known tend to be Mersenne primes (in fact, in the time since Mersenne, the largest known prime has always been a Mersenne prime); the record in early 2015 stands at: $M_{57,885,161} \approx 10^{17,425,170}$

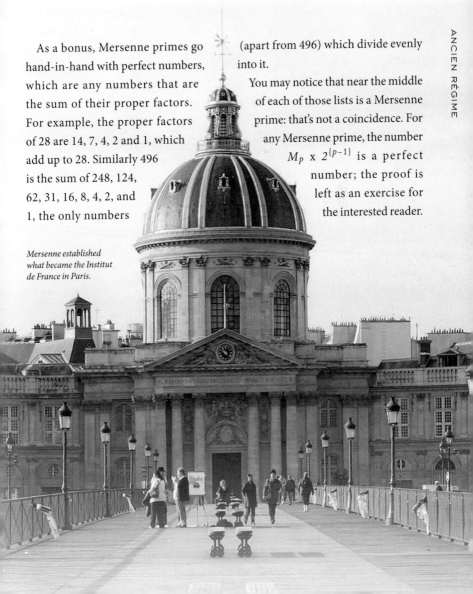

As a bonus, Mersenne primes go hand-in-hand with perfect numbers, which are any numbers that are the sum of their proper factors. For example, the proper factors of 28 are 14, 7, 4, 2 and 1, which add up to 28. Similarly 496 is the sum of 248, 124, 62, 31, 16, 8, 4, 2, and 1, the only numbers (apart from 496) which divide evenly into it.

You may notice that near the middle of each of those lists is a Mersenne prime: that's not a coincidence. For any Mersenne prime, the number $M_p \times 2^{\{p-1\}}$ is a perfect number; the proof is left as an exercise for the interested reader.

Mersenne established what became the Institut de France in Paris.

Dividing the winnings of two gamblers was a problem only mathematicians could solve.

THE PROBLEM OF THE POINTS

Two Parisian noblemen take part in a game of chance: they agree to play "best of seven."

After the fourth game, the first player, let's call him Jacques, realizes with a jolt that he has an urgent appointment elsewhere and will have to abandon the game. "But how shall we split the stakes?"

The second player, Jules, says, "It would be dishonorable to expect you to forfeit the remaining games, after all, you're ahead by three games to one; you clearly deserve a higher proportion of the pot than me."

This problem of the points was a long-standing puzzle. Luca Pacioli suggested splitting the stakes in the ratio of the current score, but this is hardly fair: if you're winning 2–0 in a first-to-10 game, you hardly have a decisive lead! A player leading 9–7 is much more likely to win than a player leading 2–0.

Tartaglia, who noticed this problem with Pacioli's solution, suggested instead that the stakes should be split according to a ratio between the lead

Pierre de Fermat's solution to the problem was logical and exhaustive.

and the number of games in the series — but this leads to the same difficulty.

Around 1654, the Chevalier de Méré asked his friend Blaise Pascal whether he could solve the puzzle. Pascal and his buddy Fermat were the first to figure out the correct answer.

They reasoned that the probability of winning depended not so much on what had already happened, but on what could still happen — if you were up 3–1 in a best of seven games match, you were as likely to win as if you were up 5–3 in a best of 11 match. In either case, you need to win one game, and your opponent needs three.

Both Pascal and Fermat came to the same conclusion, by different methods: that the stake in these games should be split in the ratio of 7:1 in favor of the leader.

Fermat did it by an elaborate system of tables, working out what results could possibly happen if the remaining games were played out.

Pascal, on the other hand, looked at it rather differently.

In mathematics, Pascal's triangle is a triangular array of the binomial coefficients.

```
              1
           1     1
        1     2     1
     1     3     3     1
   1     4     6     4     1
 1   5   10    10    5     1
```

Pascal, far more subtly, used an inductive process to decide on a fair division. This turned out, in the end, to involve Pascal's Triangle, which he didn't invent, but did write about.

Going back to our noblemen, here's the method:

- Suppose Jacques needs a more games to win and Jules needs b. The maximum number of games left is $a + b - 1$, which I'll call G.

- Look at the row of Pascal's triangle that starts "1 G" and draw a line after the bth entry. Add up the numbers on the left and the numbers on the right.

- The ratio between these numbers is the fair division of the pot between Jacques and Jules.

In the original example, Jacques (leading 3–1 in a best-of-seven match) needs one game to win; Jules needs three. There are, at most, three games left, so we need the row that starts "1 3." (The row is 1 3 3 1.) Draw a line after the third entry;

Jacques's share, on the left, is 1 + 3 + 3 = 7; Jules's share, on the right, is 1, so the pot should be split in the ratio of 7:1.

Pascal and Fermat's solution is especially significant: it's the first time expected value is used, and pretty much the birth of probability as we know it — anyone who's ever had to draw a probability tree would recognize Fermat's method of exhausting every possible outcome!

BLAISE PASCAL

One of the first serious questions to be addressed using probability was, "Should I believe in God?"

According to Blaise Pascal (1623–1662), the answer was yes, using the following reasoning: if you believe in God and it turns out you were right, you get a big win and go to heaven for eternity, while if you don't believe and it turns out you were wrong, you get a big loss and wind up in the bad fire pushing rocks about for the rest of time. By contrast, if you believe in God and there isn't one, you lose a few Sundays in the church compared to not believing in a God who doesn't exist — a very small insurance policy.

Pascal's number triangle solved the problem of the points.

79

Pascal conducted experiments with air pressure and vacuums, including measuring the pressure at the summit of the Puy de Dome, an extinct volcano in the Auvergne.

Pascal's Wager, as it's known, is the kind of thing undergraduate philosophers like to pull apart (which God? What if He only lets in nonbelievers?), but is still an interesting approach even if you don't agree with it.

Pascal, incidentally, was deeply religious, a follower of a Catholic splinter group known as the Jansenists.

He's also known for writing extensively on Pascal's Triangle, which has all sorts of applications in probability, algebra and number theory. To construct it, you start by writing the number 1 on one line, followed by two ones on the next line, either side of the 1 above it. The rule is then, on each successive row, add up the neighbors immediately diagonally above — so on the next line, between the two 1s, you'd write 2. Then put a 1 at the start and end of each row, making the second row 1 2 1.

The Sierpinski triangle forms an infinite pattern where an equilateral triangle is repeatedly divided to create more triangles.

It's followed by 1 3 3 1, 1 4 6 4 1, and so on. There are many, many patterns to be found in Pascal's Triangle, including fractals like the Sierpinski triangle, so it's worth having a play with it on a rainy day.

But wait — there's more! Pascal was also a revolutionary physicist. By climbing the Puy de Dôme with a set of mercury barometers, he managed to show that vacuums could exist, despite what Aristotle had said — the SI unit of pressure is named the pascal in his honor. He was also the inventor of mechanical calculators called Pascalines.

Oh, and he invented projective geometry, too, adding "points at infinity," making the math of perspective much more tractable (at the slight cost of having to use an extra dimension).

Pascal died aged just 39 after living a life plagued by ill-health and brushes with death.

PIERRE DE FERMAT

In 17th-century France there were three mathematical giants: Pascal, from whom we've heard; René Descartes, from whom we'll hear shortly; and Pierre de Fermat (1601–1665), from whom we'll hear now.

The one thing (nearly) everyone knows about Fermat is his celebrated "Last Theorem". After solving problem II.8 of Diophantus's *Arithmetica* (given a rational number, k, find two other nonzero rational numbers u and v such that $k^2 = u^2 + v^2$), he wrote the most famous marginal comment in history. Translated from Latin, it says:

Fermat's spiral, also known as a parabolic spiral.

> It is impossible to separate a cube into two cubes, or a fourth power into two fourth powers, or in general, any power higher than the second into two like powers. I have discovered a marvellous proof of this, which this margin is too narrow to contain.

Or, in other words, there are no integer solutions to $a^n + b^n = c^n$ for $n > 2$. Now, it's clear that Fermat had solid proofs for $n = 3$ and $n = 4$, which is a good start — but there's no evidence he had a proof, marvellous or otherwise, for higher powers, and it's a fair guess that whatever he'd come up with was mistaken — or else it would certainly have been rediscovered in the 350-odd years before Andrew Wiles finally cracked it.

There was more to Fermat than

Pierre de Fermat on a copper engraving by François de Poilly the Elder.

a cryptic and probably erroneous comment in a textbook, though. He was a supremely gifted amateur mathematician (in his day job, he was a lawyer) who laid some of the foundations for calculus.

His idea of *adequality* isn't all that far removed from the principles behind nonstandard calculus — as well as the important proof technique of *infinite descent*, which relies on being able to say "if an integer solution to this exists, then a smaller integer solution must also exist — but eventually you run out of smaller numbers, which means there can't be a solution."

Fermat disagreed with Diophantus's approach to his puzzles that any solution (involving fractions if necessary) was fine, and instead insisted on finding all of the possible solutions — and using only integers.

Fermat studied factorization (he has a method named after him, as well as Fermat's Little Theorem: for any prime p and any integer a, the number $a^p - a$ is a multiple of p), prime numbers (a particular class of prime loosely related to the Mersenne primes are known as the Fermat primes), physics (he used the idea that light travels by the quickest path to derive Snell's Law in 1657), and, as we've seen, probability. Fermat can claim to have performed the first rigorous

THE CHEVALIER DE MÉRÉ'S PROBLEMS

Problem 1: Roll four dice. If you throw at least one six, you win; if not, you lose. Are your chances of winning better than even, worse than even, or exactly even?

Problem 2: Roll a pair of dice 24 times. If at any point you roll a double six, you win; otherwise you lose. Again, are your odds better or worse than even?

These two problems were the ones Fermat analyzed: he reasoned that the probability of losing in the first case was the probability of each die not showing a six — which works out to be $(^5/_6)^4 \approx 48.23\%$ or so, meaning the chance of winning is a little over even. A similar approach deals with the second question. The probability of losing is $(^{35}/_{36})^{24} \approx 50.86\%$, so your odds are worse than even.

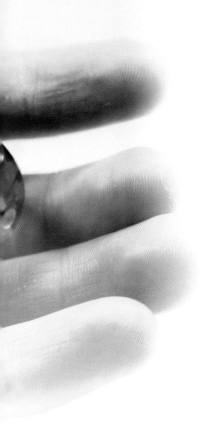

What are the chances of rolling one six with four dice?

statistical analysis in working out another pair of problems set by the Chevalier de Méré.

What are the chances of rolling a double six with two dice?

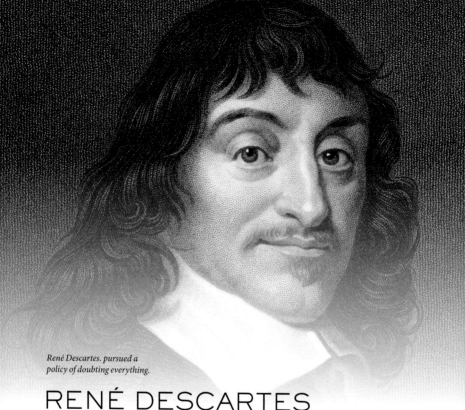

René Descartes. pursued a policy of doubting everything.

RENÉ DESCARTES

Mention René Descartes (1596–1650) to most people, and (if they've heard of him), they'll say, "I think, therefore I am."

Descartes is best known as a philosopher, although his influence spreads far wider. For example, his doctrine of doubting everything that can possibly be doubted is the bedrock of modern science.

But he isn't remembered just for his influence on scientific methodology.

He was an able scientist in his own right, rediscovering Snell's Law which, in fact, was actually discovered by Ibn Sahl, 600 years before Snellius. In France, however, it's known as Descartes's Law.

But, of course, it's his mathematical contribution we're interested in. The story goes that René was trying to get to sleep, watching a fly buzz about the room, wondering idly how one might describe its motion.

He realized that if he called the distance it was from one wall of the room x, the distance from a perpendicular wall y and the distance from the floor z, the motion of the fly was a curve he could describe geometrically *and* algebraically. He was the first to make this link, making him the founder of analytical geometry — and x, y and z are called *Cartesian coordinates* after him. Having established this link, he then did the unthinkable and suggested that things like x^4 were worth thinking about, even if they didn't represent physical objects.

Descartes established the convention of calling numbers you know a, b and c, and numbers you don't x, y and z, as well as the notation for expressing powers, laid a few foundations for calculus and, just as a by-the-way, was the first to use a form of the diagonalization argument Cantor would borrow some way down the road.

> "With me, everything turns into mathematics."
>
> René Descartes

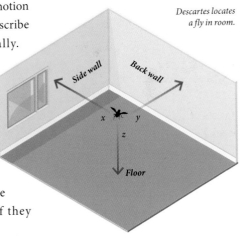

Descartes locates a fly in room.

Side wall

Back wall

x y

z

Floor

CHAPTER 4
THE SHOULDERS
OF GIANTS

Math is used to solve some of the mysteries of the universe.

"Nature and Nature's laws lay hid in night; God said,
'Let Newton be!' and all was light."

Alexander Pope

THE BIG SECRET

How far away is the Moon? How big is it? What about the Sun? That bright thing that looks like a star, but wanders around the sky, is that really a star?

How big is the Earth? What shape is it? Why are there seasons? Why does the Sun move like that?

If you imagine knowing nothing about space, you can ask any number of questions that really aren't obvious to everybody, starting with the ones above, and you'd probably get bored thinking up questions before you ran out of things to ask.

"But, but, but!" I can hear the mathematicians protesting already,

Much of mathematics was developed to help us understand the solar system.

"That's astronomy! Or at best, physics. How can you sully the purity of our immaculate subject with this reality-centric nonsense?"

Well, my imaginary mathematical friends, let me tell you math's Great Big Dirty Secret: a vast amount of the math that was developed over the last three millennia or so has been developed specifically for the purpose of understanding more about our planet and our place within the Solar System. Obviously, "our planet" is a somewhat geocentric approach, so I'd like to give a quick shout-out to any aliens who've come across a copy of *The Mathematics Bible*, by way of balance.

To discuss the development of math without talking about astronomy would be like talking about biology without mentioning that it's all based on evolution — it can probably be done, but it would be a complete disservice to the subject.

Math and astronomy are intrinsically linked, as this astronomer at the U.S. Naval Observatory in Washington DC in 1925 would testify.

It's the stuff of legend that, until the Middle Ages, everyone believed the Earth was flat. Yet, in all likelihood, not very many people really thought that much about it.

In any case, the idea of a spherical planet goes back to at least the ancient Greeks (maybe 600 BCE), although the first known attempt to measure it was made by Eratosthenes around 240 BCE.

Eratosthenes's method is quite ingenious: knowing the Sun was directly overhead Swenet in Egypt where he lived (it's on the Tropic of Cancer, which means at midsummer, the Sun is straight up), he measured the angle of the shadow cast by an upright in Alexandria, which is almost exactly due north of Swenet, and found it to be a 50th of a circle. He deduced that Alexandria was 1/50 of the globe north of Swenet — and he knew how far apart the two places were!

There's some debate about how accurate Eratosthenes's measurement was — not least because of a few assumptions he made. Swenet isn't due south of Alexandria, and the distance he said was between them is a suspiciously round number. Neither do we know which measurement system he was using! Different authors use different distances for the *stade*,

Tropic of Cancer

World map with Tropic of Cancer.

and depending on which one you pick, he was right to about 2 percent, or to about 16 percent. One of these is more impressive than the other — although, if you take account of the errors and use the same method with modern distances, you get an Earth with a circumference of 24,900 miles (40,074 km), which is only a couple dozen miles off of the known answer today.

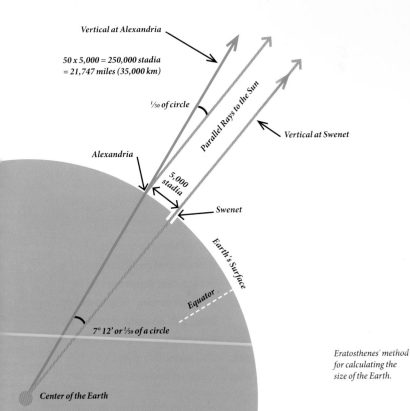

Vertical at Alexandria

50 x 5,000 = 250,000 stadia = 21,747 miles (35,000 km)

¹⁄₅₀ of circle

Parallel Rays to the Sun

Vertical at Swenet

Alexandria

5,000 stadia

Swenet

Earth's Surface

Equator

7° 12' or ¹⁄₅₀ of a circle

Center of the Earth

Eratosthenes' method for calculating the size of the Earth.

Mercury   Venus   Earth   Mars   Jupiter   Saturn   Uranus   Neptune

THE EPICYCLE SOLUTION

Although the ancients didn't have a problem with a round Earth, what they did struggle with was where the Earth sat in the universe.

It stood to reason that our planet was at the center of everything, didn't it? The only problem was that the other planets didn't really behave like it as they wandered across the sky.

Instead of traveling in nice, straight lines at a more-or-less constant speed, they would routinely speed up and slow down, and even go backward. Astrologers will tell you that there is "huge significance" to this kind of retrograde moment, and blame their inability to find their car keys on Mercury being retrograde rather than their own disorganization. But I digress.

About 300 BCE, Apollonius of Perga

came up with a method to model these changes in motion, called epicycles. Each planet, he reasoned, traveled largely around an enormous circle (the *deferent*) in roughly the same plane as the Sun; around its position on the deferent, it would orbit in a small cycle — an *epicycle*.

This did a decent job of explaining the Sun's motion, although around 200 CE when Ptolemy tried to reconcile it with the actual data gathered by the Babylonians, he hit a problem: the planets weren't revolving at a constant rate.

That said, if he measured the angle from somewhere else, a place he called "the *equant*," he *could* make it work. In fact, his predictions for the motion of Jupiter were still workable in the 1500s.

In the 12th century, an epicycle-free theory came to light courtesy of Ibn Bajjah, but epicycles remained the basis of Western astronomy until the time of Kepler — of whom more shortly.

In fact, any reasonable path across the sky can be plotted by means of epicycles — effectively using a Fourier series. If your model doesn't fit the path of a planet, you can simply add more epicycles until it does — to the point where "adding epicycles" has become a scientific insult.

By adding epicycles, you make an inelegant model even ineleganter (well, it ought to be a word — it's quite self-descriptive), when there's an underlying model that's simpler and better.

Ptolemy realized that the planets did not revolve at a constant rate.

AND YET IT MOVES

Although Nicolaus Copernicus (1473–1543) added epicycles to Ptolemy's method, he did make one giant breakthrough.

Copernicus monument in Warsaw, Poland.

Copernicus noticed that if, instead of picking the Earth as the center of planetary motion, you picked the Sun, all of the sums became substantially easier.

His colleagues urged him to publish his work, but he declined, recognizing that moving the center of the universe was likely to generate controversy and a backlash from both the community of astronomers and the Church. Actually, the Church didn't seem too worried. When the pope's secretary (Johann Widmannstetter) lectured in Rome on the heliocentric theory, Pope Clement VII and several cardinals showed up, one of them (the Archbishop of Capua) adding his voice to the "you ought to publish this stuff" crowd. When he eventually

Kepler hit on the idea that the planets might move in ellipses rather than circles.

until a couple of decades afterward. It was Kepler who finally got rid of the epicycles.

The problem with Copernicus's system, he held, was that it insisted that everything moved in *circles*. If, instead, you allowed planets to move in *ellipses*, with the Sun at one focus, you could describe everything in an even *more* straightforward manner.

This idea is Kepler's first law. The second says that if you draw a line from a planet to the Sun at several regular points in time, the area in each section is equal; the third is that the orbital period of a planet is proportional to the size of the ellipse (strictly, the length of its semi-major axis) to the power of $\frac{3}{2}$.

Newton eventually showed why these laws were true, using the fact that the force between two bodies varies inversely with the square of the distance between them.

Kepler and Copernicus managed to avoid the wrath of the Church in proposing their heliocentric universes, but Galileo wasn't quite so fortunate.

published, he dedicated it to Pope Paul III.

It took a while for the Copernican system to catch on. His masterpiece, *On The Revolutions Of Celestial Spheres*, was published in the year of his death (the final draft said to have been brought to him on his deathbed), but its two major champions — Galileo Galilei (1564–1642) and Johannes Kepler (1571–1630) — weren't even born

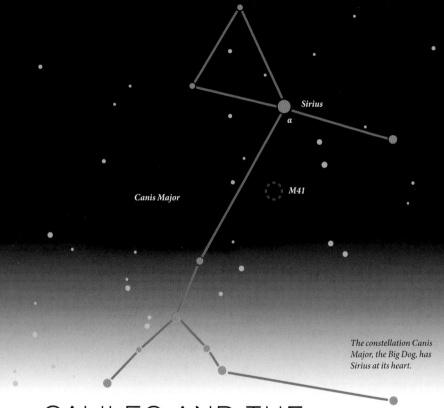

Sirius

α

Canis Major

M41

The constellation Canis Major, the Big Dog, has Sirius at its heart.

GALILEO AND THE STELLAR PARALLAX

One of the reasons for the average astronomer at the turn of the 17th century having doubts about the Sun-centered theory of the planets was, on the face of it, reasonable.

The argument was, if the Earth went around the Sun, you ought to be able to see a pronounced shift in the positions of the stars between, say, midsummer and midwinter. However, despite looking really hard, nobody had seen any evidence of such a shift, called the *stellar parallax*.

Yet the stellar parallax is a real phenomenon: it's just really hard to detect. The reason it couldn't be seen in Galileo's time is that it's extremely small. If you take my favorite star, Sirius, which is about 8.6 light years away, and compare that distance with the diameter of the Earth's orbit — about 16 light minutes — the distance to the star absolutely dwarfs how far the Earth moves (it's more than 28,000 times as large). The distance Sirius would appear to move across the sky is too small for the telescopes of the day to notice.

This argument was quite enough for the Roman Inquisition, who, in 1615, decided that heliocentrism wasn't just wrong, it was "foolish and absurd in philosophy, and formally heretical," and forbade Galileo from advocating his viewpoint. Galileo then did about the craziest thing he could, which was to write a book as a dialogue between Salviati (who proposes Galileo's views), Sagredo (a neutral layman) and Simplicio (an advocate of the Ptolemaic system).

Simplicio borrowed many of his arguments from the Pope, who felt like he'd been made to look a fool. The Pope wasn't pleased: Galileo was convicted of heresy, living his last 9 years under house arrest, and his best-seller placed on the index of banned books. It was 1835 before Galileo's works were allowed in print again.

The inquisition demanded that Galileo stop believing in this nonsense — that he "recant, curse and detest" these opinions. Legend has it that after cursing them, he muttered mischievously: "And yet, it moves!" Unlike most legends, this appears to be at least contemporary.

"And yet, it moves!" Galileo Galilei following his Inquisition hearing.

WHAT GALILEO DID NEXT

So, Copernicus said, "Put the Sun in the middle and everything gets easier."
And Kepler said, "How about ellipses rather than circles?"

And Newton said. "Here we go, apply a bit of this new calculus malarkey to it and boom, the sums all work out." That was that for the math of astronomy, right?

Not exactly, no.

Aside from upsetting the Pope, Galileo had a few more tricks up his sleeve.

Galileo saw sunspots through his telescope.

He provided further support for the heliocentric model by spotting a few of Jupiter's moons with a telescope he'd just invented — he initially thought they were stars, but after a few days' observation realized they were moving around Jupiter, which made no sense in a geocentric universe.

He observed sunspots, contradicting Aristotle's view of the Sun as perfect and unchanging.

Also contrary to Aristotle's assertions, he showed that the rate two bodies fall at is independent of their mass; that is, if you drop a marble and a cannonball at the same time from the same height, they'll both hit the floor at the same time (ignoring air resistance).

He came up with Newton's First Law of Motion (objects in motion keep going at the same rate, unless something slows them down).

He pre-empted Georg Cantor by showing there were as many square numbers as there are whole numbers (because every number has a square, there must be as many — even though square numbers appear much more sparsely than whole numbers!).

He also said that the laws of physics

Galileo pre-empted Newton's work on gravity and the Laws of Motion.

would be the same in any system moving at a constant speed in a straight line.

That's why, if you're on a train and jump straight up in the air, you land at the same point in the carriage as you took off from, rather than hitting the back of the carriage.

This idea? It's the basic principle of *relativity* — and it wouldn't be properly developed for another 300 years.

EINSTEIN SHEDS A LITTLE LIGHT

Sunlight (the cliché says) is the best disinfectant. Light itself, though, was what eventually undermined Newton's explanation of the heavens.

Not everything obeys Sir Isaac Newton's laws.

Let me be perfectly clear: Newton's Laws are an excellent model for how things behave in most circumstances. They're not *wrong* — at least, when applied within the domain where they apply. Where they *don't* apply, though, is when things move very fast. Or if they're massive .

Which is where Albert Einstein (1879–1955) comes in. Famously, his experiments were generally done in his head rather than in the laboratory, but it's hard to think of a laboratory that had more far-reaching results. He started from two postulates and developed a theory that revolutionized physics.

The first was what Galileo said: the laws of physics are the same for all observers in uniform motion relative to each other. The second was that the speed of light is the same for everyone, no matter how they or the light are moving.

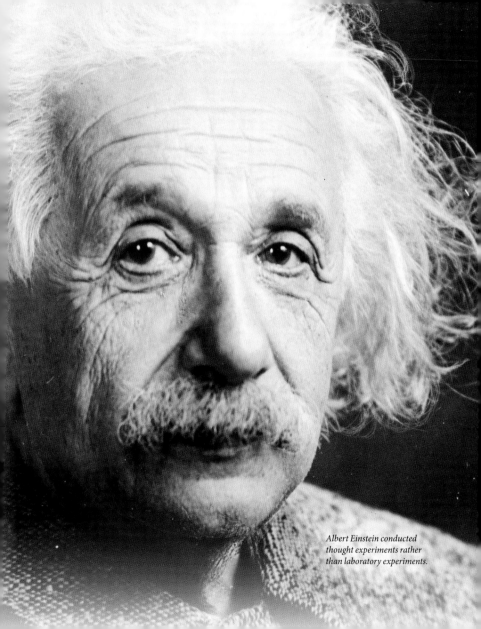

Albert Einstein conducted thought experiments rather than laboratory experiments.

Flying in a star cruiser at close to the speed of light will have various effects on time, and the appearance of the star cruiser.

There are a lot of consequences from that, the main mind-bending one being that time doesn't flow in a nice straight line for everyone. If I fly to the moon and back at close to light speed, my clock will be slightly out of synchronization with yours when I return.

Things that look simultaneous to you may not look simultaneous to me, depending on how I'm moving. Nothing (at least, no information and no physical object) can travel faster than the speed of light. Objects moving may be shorter than they appear.

And, more famously than anything else: the energy contained in an object's matter is equivalent to its mass, multiplied by the square of the speed of light. Or rather:

$$E = mc^2$$

And that's just special relativity. Einstein later proposed his theory of

Einstein's most famous equation.

general relativity, which reformulates gravity as the curvature of space-time, a four-dimensional extension of space which includes, you've guessed it, time.

I can hear objections rising up all over the place again: "But this is definitely *physics*!" It is that, but there's a reason I've included it: in most of this chapter, math has ridden in on its shining white charger to help poor models agree better with experiments.

Einstein's work is different. He came up with the equations, waiting for the experiments to prove them correct.

Astronomers have a lovely job. They get to look at the results of the great cosmic computer program and figure out what's going on. On the other hand, mathematicians — well, we get to look at the source code.

It did not last: the Devil, shouting "Ho.
Let Einstein be," restored the status quo.
                                   J. C. Squire

CHAPTER 5
THE INFINITESIMAL

A tortoise outruns a hare, a circle is carefully trapped between polygons, two of the greatest minds in history get into a "No, I thought of it first" verbal boxing match, and we are able to measure slopes and areas.

Zeno's Achilles and the Tortoise *was adapted to become the* Tortoise and the Hare *in later versions.*

ZENO OF ELEA

The paradox is one of the mathematician's favorite tools. Creating absurd statements from logical premises is a great way to show that there's a problem with either the premises, the logic, or your understanding of what's absurd.

Zeno of Elea (around 490–430 BCE) appears to have had no other hobbies. He was a follower of Parmenides, who held that there was no such thing as change and that motion was an illusion; it is thought that many of Parmenides's opponents had created paradoxes to undermine his teachings.

According to ancient Greek street rules, if you dissed Parmenides with paradoxes, you could expect Zeno to fling paradoxes right back at you. It was on. On like the Parthenon.

Zeno's most famous paradox involves Achilles and the Tortoise, who engage

Zeno of Elea.

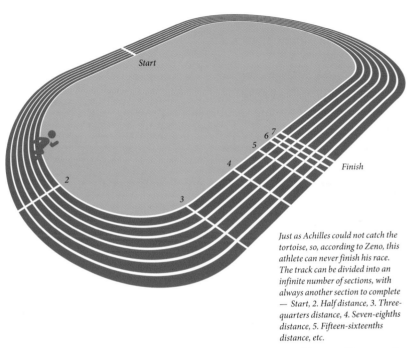

Just as Achilles could not catch the tortoise, so, according to Zeno, this athlete can never finish his race. The track can be divided into an infinite number of sections, with always another section to complete — Start, 2. Half distance, 3. Three-quarters distance, 4. Seven-eighths distance, 5. Fifteen-sixteenths distance, etc.

in a foot-race that everyone expects to be rather one-sided.

Achilles, naturally, concedes a head start to the Tortoise. After all, he is the swiftest of all men, whereas the Tortoise is among the slowest of all creatures.

Zeno agreed that the outcome was in no doubt... but he disagreed about who would win. You see, by the time Achilles catches up with where the Tortoise started from, the Tortoise has moved on to a different point.

By the time Achilles reaches that point, the Tortoise has moved on yet further — and so on! Achilles can never catch up.

Similarly, Zeno claims it's impossible for an arrow to fly through the air and hit a target: before it gets to the target, it has to go halfway. Before it gets there, it has to go a quarter of the way, and so on.

Time is infinitely indivisible, and there's never enough time to do an infinite number of tasks. (My personal rebuttal would have been to get a bow and arrow and say, "Okay, Zeno, you stand there, and I'll attempt to fire this impossible arrow at you.")

There are others: the arrow that can't move because at any given time it's frozen in space. Nothing can be in two places at once. Therefore, at a particular point in time, the arrow must be at a particular point in space. If it is in that point, at that instant, then it must be stationary and it should fall out of the sky.

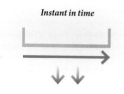

Then there's the sack of rice that makes a noise when spilt, even though each individual grain is silent. Zeno is reputed to have come up with some 40 paradoxes, although only nine survive today.

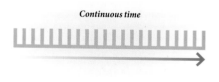

Continuous time

Arrow flies on.

Instant in time

During any single moment the arrow should be stationary and fall.

The paradoxes all have fairly elegant mathematical solutions — as well as philosophical ones. Archimedes and St. Thomas Aquinas are among many who have had a stab at them and, although practical solutions like running a race between an athlete and a tortoise are convincing, they don't explain why Zeno's argument is wrong. For that, you need to use the calculus.

How much noise does a grain of rice make?

ARCHIMEDES AND THE INFINITESIMAL

Archimedes was one of the first mathematicians to work out the area of shapes that weren't nice and regular.

It's easy enough to work out the area of a square or a rectangle; from there, a triangle is straightforward, and that opens up parallelograms, trapeziums, rhombuses, all manner of polygons, regular and not — but it doesn't really help you with curved shapes.

A parabola, for example, is roughly the shape you get if you throw a ball in the air — you might recognize it as the graph of a quadratic function. If you try to split it into triangles (or any other straight-edged shape), you very quickly notice that there's always a curvy bit left over.

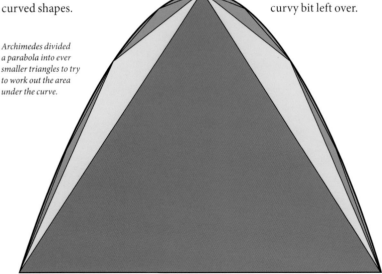

Archimedes divided a parabola into ever smaller triangles to try to work out the area under the curve.

Archimedes is said to have been working on curves and circles when he was murdered by a Roman soldier.

Archimedes, ingeniously, decided not to let that get in the way: he recognized that if you continued to add triangles, you could get closer and closer to the value of the area under the curve.

After you draw the first triangle, there are two more triangles to add immediately, one on either side. Each of these works out to be an eighth of the size of the first triangle — in all, you're adding a quarter of the area of the first triangle.

In the next step, you add on four more triangles, each an eighth of the size of the one before it; if you work through the sums again, you find that you've added a quarter of the area of the second set of triangles.

The sequence continues: at each step, you add a quarter of the area of the step before. In modern notation, if the area of the first triangle was **A**, you would get:

$$\mathbf{A} \left(1 + \tfrac{1}{4} + \tfrac{1}{16} + \tfrac{1}{64} + \dots\right)$$

Archimedes, of course, didn't have modern notation, let alone the formula for a geometric series that a good high school student would automatically reach for. Instead, he had to prove it for himself.

To do this, he considered a square of area 4 (like the one shown), putting a square of area 1 in the bottom left, a square of area ¼ above and to its right, a square of area ¹⁄₁₆ above and to the right of that, and so on forever.

You can see that one in three of the biggest squares are green, one

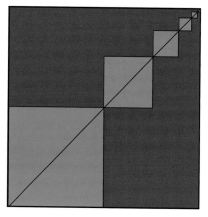

Archimedes imagined an ever-smaller series of squares.

in three of the next-biggest squares, and so on — meaning that the green squares make up ⅓ of the total area, or ⅓ of a unit: the area of the parabola is ⅓ of the area of the inscribed triangle, which is simple to work out.

WHY IS PI PI?

One of the shapes Archimedes tried to find the area of was the circle. It turns out you can't do that exactly using the methods available to the ancient Greeks, but Archimedes didn't let this put him off and had a decent attempt at it all the same, using the following method.

First of all, draw a regular hexagon whose points lie on the circle. That's easy enough to do — and you can find the perimeter of the hexagon without too much difficulty.

If the circle has a radius of 1 unit, the perimeter of the hexagon is 6.

You can also draw a regular hexagon with sides just barely grazing the edges — that turns out to have a perimeter

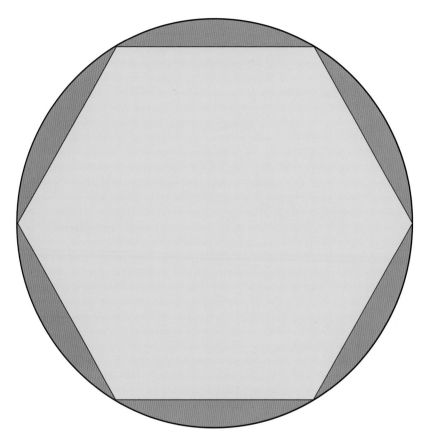

*A circle with a radius of 1 containing a hexagon
with a perimeter of 6.*

of 4√3. That means the perimeter of the circle—in this case, 2π — has to be between 6 and 6.93 or so, making π somewhere between 3 and 3.46.

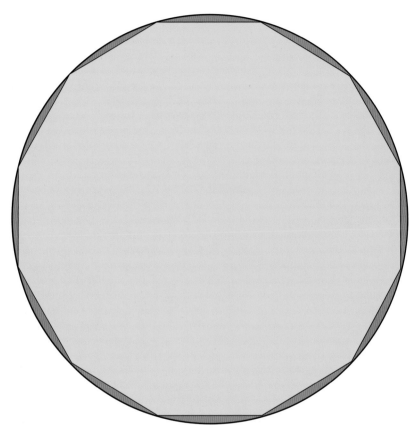

A dodecagon inside the circle took Archimedes closer.

Luckily, Archimedes didn't stop there: he then did the same thing with a dodecagon — 12 sides give a considerably rounder shape than a hexagon, making it considerably closer to a circle, giving tighter limits on the value of 2π: with this approximation, it's between 24(2–√3) and 6(√6–√2), roughly 6.43 and 6.21, making π between 3.11 and 3.21.

Next stop: a 24-gon, which is better yet. According to Conway, it's an icosikaiteragon. (I don't know any mathematician who would prefer that name to a 24-gon). Then a 48-gon and a 96-gon, which was as far as he went.

Because he was able to work out rational bounds for the irrational values coming out of his pictures (even the values for the 24-gon are tricky to express in terms of roots), he managed to put a value on π of between $^{223}/_{71}$ and $^{22}/_{7}$ — in decimals, between about 3.1408 and 3.1429.

Taking their mean gives a value for π that's off by about one part in 12,000 — not a bad effort.

His method was the state-of-the-art circle area finding method, in Europe at least, until the very end of the 17th century, although the bounds improved with the algorithms for computing trigonometric values — by 1630, π was known to 39 digits:

3.1415926535897932384 6264338327950288419

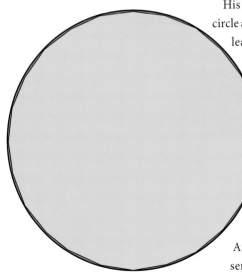

A 24-gon, or icosikaiteragon, inside a circle.

Approximations using infinite series finally beat that record in the year 1699.

LIU HUI'S METHOD

Archimedes' method was the state of the art in European mathematics until the 17th century. However, it was superseded by the Chinese mathematician Liu Hui (225–295 CE) around the 3rd century CE.

His basic method was the same as Archimedes', but he made an improvement. He noticed that the difference in areas between one inscribed polygon and the next was approximately ¼.

He used this to effectively weave gold from straw: once he had worked out an approximation to π using a 96-gon, like Archimedes, he took the difference in area between the 96-gon and the 48-gon, multiplied it by ⅓ (because, like Archimedes proved, adding up terms where each is a quarter of the one before, you end up with a third of what you started with) and tacked it on.

Zu Chongzhi took Archimedes' method to a 12,288-gon.

This meant that Liu Hui could get an accuracy equivalent to a 1536-gon using just a 192-gon! There's efficiency for you. Later mathematicians took it further still: Zu Chongzhi (429–500 CE) used a 12,288-gon to determine 3.14159261864 < π < 3.141592706934.

Their mean is correct to better than three parts in a trillion. He also used an interpolation algorithm to find the famous approximation for π of ³⁵⁵/₁₁₃ — correct to one part in 12 million.

It's hard to think of a situation where that wouldn't be accurate enough.

NEWTON
VS. LEIBNIZ

*Newton and Leibniz
could have achieved
so much more.*

*It's one of the great mathematical
what-ifs: what if, instead of spending the
last few decades of their lives bickering
over who came up with calculus first,
Newton and Leibniz had focussed on
extending it, or even reconciling the
notational differences between the two.*

The timeline isn't in dispute:
Newton started working on
calculus in 1666, but didn't immediately
publish. Leibniz, in the course of his
studies, started work on his version in
1674, and published in 1684.

Newton's *Principia* came out in 1687, and explained the geometrical form of calculus in some detail, which L'Hôpital acknowledged in 1696 when writing a textbook about Leibniz's version.

Newton didn't explain his notational variant fully until 1704, nearly 40 years after he had the idea.

There is very little dispute about who thought of it first. However, there was a more serious allegation, first made by Newton's buddy Nicolas Fatio deDuillier in 1699, that Leibniz had stolen Newton's ideas.

In 1712, the Royal Society of London (Chairman: a Mr. I. Newton, so clearly an independent arbiter) published a document entitled *Commercium Epistolicum*, purporting to show all of the relevant correspondence demonstrating this fact. Leibniz was not invited to give his side of the story.

Leibniz did make things slightly worse for himself, though. One of Newton's documents, circulated in the mid 1670s, did mysteriously turn

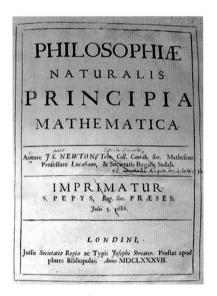

Newton's own copy of his Principia, *with hand-written corrections for the second edition.*

up among his papers. Elsewhere, documents were found to have been altered, added to, or (in one case) to have had the date changed. At certain points, he corresponded with Newton and asked his colleagues for advice.

Even if Leibniz *had* taken some inspiration from the documents in his possession (which a competent plagiarist would have disposed of properly, and

Leibniz was nothing if not competent), anyone who's ever tried to teach or learn calculus will recognize that it's hard enough to master even with a complete textbook in front of you.

To have reconstructed the entire edifice from a few hints, starting in an entirely different place (Leibniz began with integral calculus, whereas Newton was a differentiation-first man) and with a different, far superior notation (Newton denoted different derivatives with dots and dashes, whereas Leibniz used a system involving a lot of letter ds that's used almost universally now) — to me, that would be almost more of an intellectual achievement than coming up with calculus from scratch.

I feel a bit sorry for Leibniz — he got there second, as independently as I can imagine, and is still portrayed as a crook and a thief, in large part because Newton abused his position in the Royal Society to wage a campaign against him, while pretending to be neutral. I think we can all see who the villain of the piece is, here, can't we?

Sir Isaac Newton accused Gottfried Leibniz (above) of plagiarism.

9 %

HOW DIFFERENTIATION WORKS

As you approach a steep hill, you might see a warning sign that says "12%" or maybe "1:8," telling you how steep the hill is.

A 12 percent slope means you go up 12 meters for every 100 you move horizontally; 1:8 means you go up one meter for every 8. In both driving and in math, this is known as the gradient of the slope.

There's a difficulty, though: what if

the hill gets gradually steeper? How do you account for a changing gradient? That's where calculus comes in.

What you would do, if you were obsessed with getting perfect accuracy at some point, would be to take a small section of the slope starting at that point, and work out the gradient of that section. How many millimeters across has it gone for every one it's gone up? You

could do the same on a micrometer scale, then a nanometer, and so on. The values would converge on a (probably not very meaningful) number.

With mathematical functions, the number is slightly more useful — but the technique is precisely the same: you work out the value of the function at the point you're interested in. Let's call that $f(X)$. You'd then work it out a small distance further on: $f(X+h)$. To find the gradient of the line between those points, you need to divide the difference between these (how far up you've gone)

by how far across you've moved — h. That makes the gradient:

$$\frac{f(X+h) - f(X)}{h}$$

Making h smaller and smaller gives you a better and better approximation of the gradient. By a bit of mathematical sleight of hand, it is possible to see (at least for well-behaved curves) what happens when h becomes zero — which is the mathematically-defined gradient at that precise point.

SIR ISAAC NEWTON

I think it's fair to say that Isaac Newton (1643–1727) was a pretty influential chap. He is credited not so much with the discovery of gravity (people had, after all, noticed things falling to the ground before), but with figuring out how it fit into a mathematical scheme.

Classical mechanics is often called Newtonian mechanics, in his honor, as a result of his work. He made significant advances in optics (that Pink Floyd album cover with the prism? Newton did that first, although his influence on progressive rock is unrecorded).

And, most famously, he developed calculus, before spending decades squabbling with Leibniz about priority.

The mathematical framework for gravity was an especially big deal. He managed to account for Kepler's laws of planetary motion, then went on to tackle tides, comets and other theoretical problems in astronomy. He dealt with a big practical one, too, building the first working reflecting telescope.

The incident with the apple? Well, he did mention taking inspiration from watching an apple fall. It made him realize that gravity must make the apple fall toward the center of the Earth and that it must pull the Earth

Sir Isaac Newton experimenting with sunlight and optics.

toward him, or it made him realize that gravity must extend above the surface of the Earth — and if so, why not to the Moon?

Mathematically, he worked out an exponential law for cooling (now, bizarrely, called Newton cooling), extended the binomial theorem so it didn't just work with positive whole numbers, developed a method for numerically solving equations (the Newton-Raphson method), and generally made progress wherever he dabbled — except, of course, in alchemy: after his death, mercury was found in his hair, which could explain his eccentricity as an old man.

Newton certainly didn't set the world of parliamentary politics alight. As Member of Parliament for Cambridge University, it is said that his only contribution was to grumble about a draft.

He was rather more effective as Master of the Royal Mint, where he waged an almost fanatical campaign against

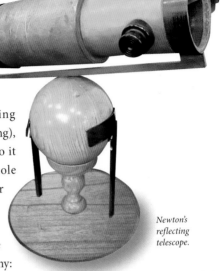

Newton's reflecting telescope.

counterfeiting, wandering around the bars and taverns in disguise to gather evidence enough to convict almost 30 counterfeiters.

He was knighted by Queen Anne in 1705, perhaps more for political reasons than in recognition of his scientific achievements, and died in his sleep in the confusing winter of 1726/27, while the old and new-style calendars were both in operation.

Leibniz's calculating machine could perform addition, subtraction, multiplication and division.

GOTTFRIED LEIBNIZ

Gottfried Leibniz (1646–1716) wasn't the first to develop calculus, but he (most likely) did it independently of Newton. More to the point, he did it better.

Aside from coming up with one of the most significant advances in mathematics ever, he has an important place in the development of calculators, which I'm sure more students will thank him for.

Not only did he invent several, but he was one of the first to figure out the binary number system, without which I wouldn't be typing a book on a computer,

and you wouldn't be leaving nice reviews of it using your mobile phone or tablet.

Leibniz's father was a professor of Moral Philosophy, and left Gottfried his library; the younger Leibniz took full advantage, reading a much wider range of books than the average schoolboy would have had access to, and becoming fluent in Latin. He studied Philosophy at Leipzig, and after gaining a Master's

Leibniz paved the way for computer programs.

degree, moved into law. Graduating from Altdorf in 1666, he eventually became a diplomat. Posted to Paris, he met Huygens and realized his knowledge of math and physics was on the patchy side. Using Huygens as a mentor, he quickly got the hang of things and, it seems, invented calculus while he was studying. Different times, indeed.

He also came up with Gaussian elimination (a method for solving multiple simultaneous equations), and Boolean logic (another thing to which we owe the computer), as well as introducing the idea of conservation of energy and anticipating Einstein by suggesting that space and time were relative rather than absolute.

He continued his work as a diplomat and legist under various patrons until his death in 1716. By this time, he was out of favor with the court and lay buried in an unmarked grave for 50 years. It seems a pity that someone who outdid Newton in many respects is rarely mentioned except alongside him.

Natural phenomenon, such as the logarithmic spiral of the nautilus shell, can be explored using infinitesimals.

NONSTANDARD CALCULUS

The traditional formulation of calculus using tangents and limits isn't the only sensible way to do things: another option is to use the idea of infinitesimals.

These are incredibly tiny numbers, much smaller than any real number. There are hints of this in earlier work, including Newton's and Leibniz's, but the limit approach eventually became the standard — until Abraham Robinson put it on a firmer footing in the 1960s.

The *hyperreal* numbers are perfectly consistent with the standard axioms of mathematics and permit a slightly more intuitive way of dealing with derivatives. There are no limits, none of the epsilons and deltas that plague standard calculus, just evaluating the function at one value and at a different value infinitesimally far away — divide by the infinitesimal, ignore any remaining infinitesimals and boom! There's your derivative.

Although a few experiments have hinted that calculus done this way is much easier to understand and learn than traditional calculus, without losing any rigor, it hasn't caught on in schools and universities … yet. One of the first non-standard calculus textbooks has been made available online (*http://www.math. wisc.edu/~keisler/calc.html*) under Creative Commons, and is worth a read.

Real numbers mean that the arrow will reach its target.

… AND BACK TO ZENO

Calculus solved Zeno's paradoxes by providing a framework for talking about things that happen on a very small scale.

Even though every time the tortoise has moved on, Achilles has to catch up, the time it takes him to do so gets shorter — and the infinitely many short times he takes to do it add up to a real number, at which point he draws level and pulls away. A similar argument deals with the infinitely many divisions of the room the arrow has to cross before it pins Zeno to the wall.

CHAPTER 6
THE FRENCH REVOLUTION

It is discovered that powers of 10 make for easy measurement conversions, a political firebrand is shot dead after a night of hard math, and infinitely many waves are added together.

A set of dueling pistols, with tools, accessories and ammunition, of the sort that brought about the demise of Évariste Galois.

DECIMALIZATION

Early on in Harry Potter and the Philosopher's Stone, *Hagrid explains the wizarding monetary system to Harry:*

> *The gold ones are Galleons.*
> *Seventeen silver Sickles to a Galleon*
> *and twenty-nine Knuts to a Sickle*
> *— it's easy enough.*

How quaint! You'd never catch a major economic powerhouse like the United States using, say, a measurement system with 12 inches to the foot, 3 feet to the yard, 22 yards to the chain, 10 chains to the furlong and 8 furlongs to the mile; now, would you?

Nor something in the UK as crazy as 16 ounces to the pound, 14 pounds to the stone, 8 stones to the hundredweight (which is nothing to do with 100), and 20 hundredweights to the ton.

It'd be still weirder if the British and American pints and gallons were different measurements, wouldn't it?

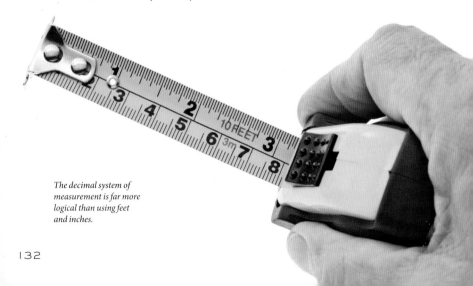

The decimal system of measurement is far more logical than using feet and inches.

What's that you say? Oh, yes, of course, that's exactly what they do, unlike every other country in the world (barring, naturally, Liberia and Burma. I have no information on which pint and gallon they use.)

This is precisely the kind of lunacy that scientists in late 18th-century France wanted to put an end to.

In those days, it was worse: different cities had different measurement systems, never mind countries.

The idea was to standardize things around a system based on units of 10, which would make everything easier to count — we have 10 fingers, after all.

Despite this extremely logical approach, the switch to decimalization didn't go as smoothly as you might think. The many governments France had at the time were really very keen on the idea, but it was deeply unpopular among the actual populace, and barely lasted a decade as the official system.

Even the precise quantities in imperial measurements are not always universal, the US gallon being smaller than the imperial gallon.

It was, however, readopted in 1837, this time for good, and scientists quickly realized that using powers of 10 made counting everything much easier. Calculations were far simpler and less time consuming, so why not make everything decimal?

Have you ever wondered what the DRG button on a calculator is for?

Decimalization didn't just affect the measurements we use today — the lovely meter and kilogram, for example. Revolutionary France also introduced decimal time and decimal angles. Have you ever wondered what the "G"

in "DRG" on your calculator stands for? It's grades or gons, of which there are 400 in a circle. The one that really stuck, however, was decimal currency.

The French franc, comprising 100 centimes, remained the currency in France until the introduction of the euro in 2002. Like every currency now in use, the euro is also decimalized.

WHY THE METER?

The original definition of the meter was brilliantly simple: it was defined so that the distance from the North Pole to the Equator, through Paris, would be 10,000,000 meters — a very nice, round number, which was around the same length as other common measures of the time, such as the yard. According to Wolfram Alpha, the distance is now 9,985,000 meters or so — within 0.15 percent of the original definition.

From the meter, it was possible to derive several other measurements: the mass of a cube of water, one meter on each side, was defined to be one metric ton — although that was slightly ungainly for everyday use; one one-thousandth of a cubic meter was defined to be a liter, with a mass of a kilogram.

The meter is now defined to be "the distance traveled by light in 1/299,792,458 of a second," which is altogether simpler, don't you think? However, it does have several benefits: it's much easier to measure accurately, it's the same everywhere and it's based on a fundamental universal constant. So — in metrology terms, at least — it wouldn't matter if the Earth was destroyed.

The kilogram, though, is now defined as "the mass of a particular lump of metal in Paris." There are proposals to replace it with something based on a fundamental constant — but it'll be a while before the definition changes.

A replica of the prototype of the kilogram at the Cité des Sciences et de l'Industrie, Paris, France.

Roman numerals make the 12-hour system seem even more antiquated on this antique pocket watch.

DECIMAL TIME

Although most of the world has succumbed to the logic and elegance of metric measurements, there's one property that remains stubbornly undecimalized — time.

There are 60 seconds in a minute. Sixty minutes in an hour. Twenty-four hours in a day. Seven days in a week, which have little or nothing to do with the months, which vary in length between 28 and 31 days, and in one case change length depending on what year it is.

Also, around the year 1800, different countries had different opinions about the correct date. Some, like Greece, Turkey and Egypt, retained the Julian calendar until well after World War I.

Enough, thought the French; this madness must stop! Hampered slightly by the problem that the 365-and-a-bit days in a year aren't a nice power of 10, they came up with a Revolutionary Calendar: 12 months, still, each comprising three "decades" of 10 days — plus five or six days as a holiday at the end of the year, partly as a holiday, and partly to keep the seasons in sync with the calendar.

Each day was split into 10 decimal hours, more than twice the length of the ones we're used to; these were made of a hundred minutes, slightly longer than regular minutes; decimal seconds, a hundred to the decimal minute, were slightly shorter than the seconds we currently use.

Sadly, both the Revolutionary Calendar and decimal time only ever really caught on in France, and even then only for a few decades. Like the QWERTY keyboard, VHS videos and Windows software, the well-established traditional calendar and clock eventually held firm against the challenge of a better idea.

This French decimal clock shows the 10-hour day on the inner set of numbers and the two 12-hour periods of the "old" system in Roman numerals in the outer set.

JOSEPH-LOUIS LAGRANGE

Joseph-Louis Lagrange (1736–1813), or Guiseppe Lodovico Lagrangia, as he was called when he was born in Turin, now in Italy, was one of the leading lights of the decimalization movement.

He was also prominent in math in general at the dawn of the 19th century. His greatest contribution to mathematics was probably the Euler-Lagrange equations, which are used to extend regular, routine calculus into the world of functions.

Where regular calculus might tell you "where is the ball rolling fastest," Lagrange's *calculus of variations* might find the quickest route for the ball to take down the hill. Instead of solving for an optimal value, you solve for an optimal *function*.

He's also known for the Lagrange multiplier, a technique used to add constraints to a problem, and worked on many of the areas covered in this book. He laid some of the foundations

Joseph-Louis Lagrange was Director of Mathematics at the Prussian Academy of Sciences in Berlin before moving to Paris.

for Galois's group theory, studied probability and number theory, tackled interpolation and Taylor series, discovered the *Lagrangian points* (places where the gravitational forces from the Sun, Moon and Earth cancel each other out, making them ideal

Lagrange avoided the guillotine, unlike other academics and aristocrats, including King Louis XVI.

places to put satellites) … oh, and he revolutionized mechanics, too, showing that everything Newton had worked out could be produced as a consequence of the calculus of variations.

Toward the end of his career, while many of his friends and colleagues were being guillotined, Lagrange was held in the highest honor by governments of all stripes — for example, when The Terror began, all foreigners were ordered to leave France, except Lagrange, who was exempted.

Although he was making preparations to escape France, quite reasonably fearing that his head may be next on the block, he was offered the Presidency of the Commission reforming weights and measures, and is largely responsible for the eventual choices for the meter and kilogram.

He is one of 72 scientists honored with a plaque on the first stage of the Eiffel Tower. Reading through his list of achievements, I can't help but think he deserved at least two.

PIERRE-SIMON DE LAPLACE

Almost as intimidating as Lagrange's work is that of Laplace (1749–1827), sometimes known as "the French Newton," although there's no evidence that he was anything like as vindictive.

I particularly like Laplace, because he did much more to establish Bayesian probability than Bayes ever did. He was also largely responsible for changing the study of mechanics from a geometrical pursuit to something based on calculus — and it's calculus for which he's best known.

He has an equation named after him, Laplace's Equation, which generates potential fields. These are used in electromagnetism, astronomy and fluid dynamics. He also gives his name to a technique, the Laplace Transform, which can turn horrible differential equations into much less formidable algebraic ones, and a symbol — the Laplacian, Δ.

Laplace's big break came when he dropped out of theology school to become a mathematician, and presented himself to top Parisian mathematician d'Alembert. Trying to get rid of him, d'Alembert gave him a huge mechanics book and said, "Come back when you've read it." A few days later, Laplace was back. There was no way, thought d'Alembert, he could have read it that quickly — so he scornfully asked a few difficult questions from it that Laplace answered with ease.

D'Alembert quickly found him a place doing routine teaching at the École Militaire, giving him plenty of time for his research.

One last thing he's known for: Laplace's Demon is the hallmark of a deterministic universe. Laplace suggested that if an intelligence knew the mass of every particle in the universe, and the forces acting on them, he'd be able to see into the future as clearly as he saw the present.

He didn't use the word demon, though. Pity.

"Nature laughs at the difficulties of integration."

Pierre-Simon de Laplace

Laplace took up a teaching post at the École Militaire in Paris.

ÉVARISTE GALOIS

Évariste Galois was born in 1811 in Bourg-la-Reine, France. His parents were ferocious anti-monarchists (his father was the Liberal mayor of Bourg-la-Reine), a political standpoint he adopted with enthusiasm.

Around the age of 14, he got bored of being brilliant at Latin and turned to math, reading Legendre's *Elements of Geometry* "like a novel" and quickly mastering it. This didn't impress his teachers, who were presumably outclassed by their bright pupil.

This would become a common theme. He failed to get in to the École Polytechnique, France's top math school because his explanations missed out steps he felt were obvious.

The following year, shortly after his father's suicide, he failed his second attempt to get into the Polytechnique — legend has it that

Galois made the grave error of becoming an enemy to King Louis Philippe, who may have ordered his death.

he got frustrated with a dim-witted examiner and threw something at him.

His most stellar mathematical achievement was to determine conditions for a polynomial equation to be solvable, and to find a connection between the solutions. He was the first to use the word "group" mathematically, and laid the foundations for much of modern group theory. The area he worked is now called Galois Theory in his honor.

Sadly, Galois's politics and ill-temper worked against him, and he was fatally wounded in a duel, aged just 20.

There were formal rules to pistol duels, which gentlemen used to resolve disputes.

GALOIS'S LAST NIGHT

Mathematics is a little bit short on romantic heroes, so we try to make the best of the ones we do have. Head and shoulders above the average mathematical romantic hero is Évariste Galois.

The legend of Galois is that he was a political troublemaker, and the draconian monarch of the day (King Louis Philippe) decided he wanted to be rid of him.

The king's accomplices contrived a situation where young Évariste, not yet 21 years old and fresh out of prison, was manipulated into challenging someone to a duel he would surely lose.

Knowing he was bound to die the next day, and fearing his priceless insights would be lost to the world, Galois stayed up all night writing out everything he knew about group theory — punctuating his work with plaintive phrases like "There is not enough time!" and declarations to his beloved Stéphanie.

Sleep-deprived, he showed up for his duel, was shot in the stomach and died a terribly romantic death in the arms of his brother, Alfred. His last words:

> *Please don't cry, Alfred; I need all*
> *my courage to die aged 20.*

Like all good legends, this has some basis in fact. Not much, but some.

Galois certainly was a political troublemaker — he had joined the artillery of the National Guard, a fiercely Republican organization.

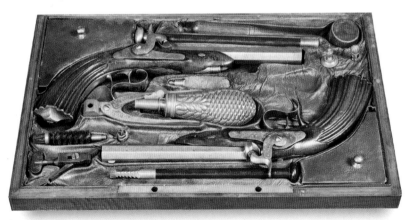

A set of 18th-century French duelling pistols with maintenance tools and accessories for making the lead ball ammunition.

He had been arrested for openly threatening the life of the king at a banquet; and he had finally been jailed for leading a protest march, heavily armed, in uniform.

It's also true that, within a couple of weeks of getting out of prison, he was shot dead, presumably in a duel. However, his fabled all-nighter is most certainly an exaggeration.

While he was languishing in prison, Galois had corresponded with the leading lights of the day and much of the work he finished up overnight was clarifying and redrafting results he'd already submitted.

Even so, the ideas were radical enough that Hermann Weyl called it "perhaps the most substantial piece of writing in the whole literature of mankind."

Duels to settle matters of honor were fought even by those in prominent positions. The Duke of Wellington fought the Earl of Winchelsea when Wellington was Prime Minister of Great Britain.

It's not clear who killed Galois, although he was certainly a man who made enemies easily. The two most likely suspects were his friends d'Herbinville, an officer of the artillery of the National Guard, and Duchatelet, who had been in prison with Galois.

However, 1832 was a time of turmoil in France (the Paris Uprising in June inspired Victor Hugo's novel, *Les Miserables*), records are scant, and we may never know who pulled the trigger.

WHAT GALOIS (AND ABEL) DID THAT WAS SO IMPORTANT

Like Galois, the Norwegian Niels Henrik Abel (1802–1829) was a genius who died early; unlike Galois,

Galois's murder went almost unnoticed during the turmoil of the 1832 Paris Uprising.

Abel's death at 26, of tuberculosis, was entirely unromantic.

The crowning achievement of his short life was to prove that no algebraic solution (solutions you can express in terms of combinations of fractions and their roots) could be found, in general, for polynomials of degree 5 or higher.

This had been an unresolved problem ever since Ferrari and his contemporaries figured out how to solve cubic and quartic equations, 250 years previously.

Aside from this, known as the Abel-Ruffini theorem, the term "abelian group" is also named for him.

An abelian group is one where the elements always commute: working out *ab* and *ba* gives you the same answer.

Galois took things a step further: he showed exactly *which* polynomials allowed for algebraic solutions.

If you can form a *solvable* (in this context, meaning you can construct it in a particular way from abelian groups) group from the possible permutations of an expression's roots, it's possible to find algebraic solutions; if not, it isn't.

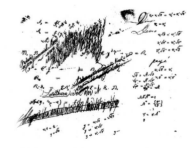

Scribbled notes made by Norwegian mathematician, Niels Henrik Abel.

Jean-Baptiste Joseph Fourier liked to bandage himself up like a mummy.

JEAN-BAPTISTE JOSEPH FOURIER

Jean-Baptiste Joseph Fourier (1768–1830), mathematician and physicist, is known mainly for the Fourier series, but had a rather odd career.

A fervent revolutionary, he was recruited by Napoleon Bonaparte as scientific advisor on his Egyptian expedition in 1798, before being pressed into service as a governor of the Isère *département*, when he'd really have preferred to be back lecturing at the École Polytechnique in Paris.

After his Egyptian adventure, he became obsessed with heat, and held it to have special healing powers. When he eventually returned to Paris, he would swathe himself like a mummy in his over-heated apartment, apparently for its health benefits. He died of heart disease in 1830.

He *should* be better known for another discovery. He worked out that the Earth was much warmer than it ought to be, given its place in the universe, and determined that the planet's atmosphere may somehow insulate it from losing heat.

Drawing on an experiment by de Saussurre, who warmed up several nested glass spheres and noted that the ones on the inside remained at a higher temperature than the outer layers, his work gave rise to the term "the greenhouse effect."

FOURIER SERIES

Fourier didn't get it *quite* right, but he got most of the way there. He believed that every function, no matter how awful, could be written as the sum of sine and cosine functions.

He missed a condition on the functions this works for; you need to be able to find the area between them and the x-axis. There are some pathological functions that don't allow this.

Napoleon Bonaparte appointed Fourier as his scientific advisor during his 1798 campaign in Egypt and Syria.

Fourier was, however, able to show his theory, somewhat informally, in his *Treatise on the Propagation of Heat in Solid Bodies* (1807) and later in *Analytical Theory of Heat* (1822).

How heat distributed itself through solids was a ... yeah, okay ... a *hot* topic at the time.

Most functions of one variable you're ever likely to come across can be written, at least in a specified domain, as an infinite sum of the form:

$$a_0 + a_1 \sin(kx) + a_2 \sin(2kx) + \ldots$$
$$+ b_1 \cos(kx) + b_2 \cos(2kx) + \ldots$$

where all of the as and bs are constants he gave a formula to find, and k is a constant that makes everything fit the domain.

This wasn't a new idea — in fact, it's slightly related to the idea of epicycles, which were used to describe the apparent motion of the planets across the sky.

Using sines and cosines to solve Fourier's favorite equation, the heat equation, wasn't a completely new idea, either. It was well-known that if the heat source varied like a sine or a cosine wave, the heat distributed through the body would do the same.

The heat equation is a *partial differential equation* — words that send shivers down the spine of all but the most dedicated of university mathematics students. They are notoriously difficult to solve, and supposedly the inspiration for one of the most famous quotes in math, by John von Neumann, who told a student, "Young man: in mathematics, you don't *understand* things; you just get used to them."

On the plus side, it's a linear equation, which means the parts it's made from aren't overly complicated (there's a more technical definition you can look up if you really want to), and linear equations have one very nice property. If you can find two solutions to them, adding multiples of them together makes a third. These are called *linear combinations* of the solutions, hence the name.

John Von Neumann advised one student that not everything in mathematics was there to be understood.

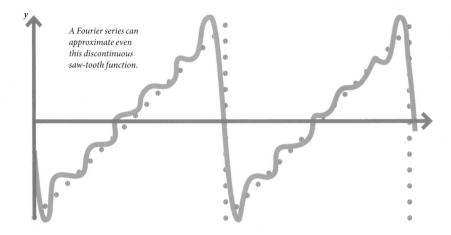

A Fourier series can approximate even this discontinuous saw-tooth function.

Fourier's insight was to say, "If we can split whatever function we happen to have up as a series of sines and cosines, we can find the solutions for each of those. Because the heat equation is linear, we can simply add up all of the solutions and get the right answer!" And, what do you know? It worked!

The sine and cosine functions behave surprisingly well when you multiply and integrate them over full cycles, and Fourier exploited this fact to come up with a method for finding out the values of the as and bs needed — the

amplitudes of each wave.

Even a strange-looking function like the one pictured, the saw-tooth function, can be approximated by a Fourier series.

The results from working out the first few terms for the sine expansion are shown — you can see that the wavy functions gradually get closer and closer to the very jagged, definitely not sinusoidal, not even continuous (a *continuous* function doesn't have any gaps in it — as usual, there's a more formal definition you can look up), saw-tooth.

American scientist Josiah Willard Gibbs (1839–1903) gave his name to an oddity of the Fourier series.

One of the oddities of the Fourier series is that near a discontinuity the approximation can go a little haywire — I know you shouldn't anthropomorphize mathematical objects (they don't like it), but it's hard not to see the function as getting confused and going slightly wobbly.

This "ringing" is known as the Gibbs phenomenon, and can be reduced by a clever (but slightly inelegant) correction called the Lanczos sigma factor.

MUSIC AND MATH — WAVES

One of the upshots of Fourier's work was the realization that you can reproduce musical notes using sine and cosine waves of different frequencies.

Recreating music mathematically allows us to record digitally.

If you take, say, middle C on the piano, the sound can be reproduced by adding together (infinitely many) sine and cosine waves of different frequencies. A decent approximation can be worked out by just taking the waves with the biggest amplitude.

This isn't restricted to middle C; it also works for chords, and by extension for entire bands and orchestras. The whole of the sound of the Vienna Philharmonic at a given moment can be represented as the sum of sine and cosine waves. The same goes for a moment of Pink Floyd's *Dark Side of the Moon* or Mannion's *Symphony for Twelve Vacuum Cleaners*.

Why is this important? It means you can preserve sound digitally. By sampling the frequencies of a sound, however complicated, sufficiently often, and making a note of the amplitude of the waves in the Fourier series,

Algorithms that could digitally preserve music were developed in the 1960s.

it's possible to recreate the sound at another time and place — which is the essence of music recording.

The more frequently you sample the noise, and the more amplitudes you keep track of, the more faithful your sound will be to the original — but in most cases, the limitation on reproducing the sound of the Vienna Philharmonic perfectly is the sound system rather than the recording.

The "Fast Fourier Transform" is the key tool. Although some of the ideas were around as early as, well, Fourier, it wasn't until the mid-1960s that algorithms started to be developed that could digitally preserve music and — more to the point — take the preserved digits and turn them back into music!

MULTICOLORED NOISE

You've heard of white noise, of course? It's a sort of a shhh sound, although there's a more technical definition: the power (related to the amplitudes in the Fourier transform) — or loudness — at each frequency is the same. But white is not the only color.

There's also pink noise, where the power is inversely proportional to the frequency. It's called pink noise because light with this sort of transform looks pinkish.

it sounds equally loud at all frequencies — because of the way brains and ears work together, we hear certain sounds more loudly than others.

MUSIC AND MATH — SCALES

The Circle of Fifths is a rather pleasing musical trick for a mathematician. First, pick a note. I'm going for C because it's my initial. Go up by an interval of a fifth (seven semitones), and you get to G. Go up another perfect fifth: D. Keep going: A, then E, then B, then F sharp, then C sharp, G sharp, E flat, B flat, F and back to C. You hit every named note, and end up (modulo a few octaves) back where you started.

Robert Brown (1773–1858), after whom Brownian Noise is named.

Then there's red noise — confusingly, also called Brownian noise after Robert Brown — where the power is inversely proportional to the square of the frequency, meaning that lower frequencies are more prominent.

Finally, there's grey noise, which is pink noise that's been adjusted so that

Or do you? There's a small problem: changing the pitch of a note by a perfect fifth means you increase its frequency by 50 percent. You can work out the effect of doing this 12 times — the frequency ends up increasing by a factor of

$$1.5^{12} \approx 129.75$$

But going up by an octave doubles the frequency, so at the very least, this ought to be a whole number — and really, a power of two!

In fact, going up seven octaves increases the frequency of the note by a factor of 128 — so the perfect fifths are close but not, as their name suggests, perfect.

To get around this little problem, fifths are adjusted to be very slightly less than the 3:2 frequency ratio one might like — it's really more like 1.498:1.

All musicians, whether they realize it or not, have a practical use for some fairly complicated mathematics.

CHAPTER 7
POWERS AND LOGARITHMS

There are many tellings of the wise man and the chessboard, none of which are likely to be true. The one that follows is as fictional as the rest.

The chess board has always held a special fascination for mathematicians.

THE WISE MAN
AND THE CHESSBOARD

The Wise Man had just finished demonstrating his invention — the game of chess — to the Shah of Persia, who was duly impressed.

He clapped his hands gleefully, immediately seeing the myriad possibilities, strategic depth and the evolution of the Elo rating system. "A marvellous game!" he said. "You must, of course, be rewarded — how may we pay you?"

The Wise Man grinned. A more perceptive Shah would have spotted the cunning glint in his eyes. However, Shahs get to be Shahs because of their parents, not because of their powers of perception.

"Sire, I am a humble man of few needs.

A small quantity of rice would suffice! If you arrange for one grain of rice to be placed on the first square of the chessboard, two grains on the second, four grains on the third, simply doubling each time… I'm sure that would be reward enough for any man."

The Shah laughed, delighted — a few grains of rice, in exchange for the greatest game ever invented? What a bargain. They shook hands, and then the Wise Man headed for the door.

"Er, excuse me, sire?"

Chess originated in China or possibly India around 1500 years ago, slowly spreading west through Persia (now Iran).

whispered one of the Shah's servants. He wasn't quite sure how to say this. "Steve, the servant over there," he pointed, "he's sent me to tell you something about that rice."

"Oh, yes, the rice. I presume it's bad news?" The Shah was perhaps more perceptive than I gave him credit for.

"Indeed, sire. Steve over there worked out exactly how much rice the Wise Man asked for, and he says it's about nine quintillion grains of rice."

"Nonsense, I've been there, and they don't grow rice in Quintillia."

"No, sire, it's a technical term for a number that I think Steve over there may

have invented purely for this situation. It's a billion billion."

"It sounds like a lot."

"It is a lot. Steve over there says, one portion of rice is about 5,000 grains of rice, so that amount would feed a population of one million for about five million years."

"It is a lot. Perhaps you or Steve could fetch me back the Wise Man?"

And, when the Wise Man returned, a gleeful grin on his face, the Shah ordered that the Wise Man's head be cut off and his teeth turned into chess pieces; even in those days, nobody liked a smarty-pants.

The Shah would have needed a pretty big chess board.

JOHN NAPIER

I don't wish for much in life. I want to be around my healthy family, have a few good students, and the time to play with good math problems

John Napier's memorial stone in St Cuthbert's Church in Edinburgh, Scotland.

Oh, and I want a nickname as good as that bestowed on John Napier (1550–1617) — Marvellous Merchiston. Napier, like Mersenne, was more of a theologian than a mathematician, but he was responsible for three large advances in 16th-century math all the same: he invented "Napier's bones," he took decimal notation and made it workable, and, perhaps most importantly, he invented the logarithm.

My primary school had a box of Napier's bones, which nobody knew how to use. Part of me wants to go back there and show everyone — they're a series of rods with times tables marked on them. If you arrange them appropriately, you can work out multiplications without having to do all the tedious learning of tables. With a bit of work, you can use them to divide and extract square roots, as well — using methods not dissimilar to the "traditional" ones, only with slightly fewer opportunities for error.

The logarithm, though, is where he really picked up the ball and ran with it. In the hierarchy of mathematical operations, you go from

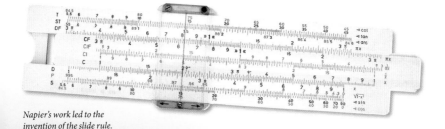

Napier's work led to the invention of the slide rule.

counting up and down (really easy), to adding and subtracting (harder to do by hand), to multiplication and division (harder still) to powers and roots (generally, quite hard).

For example, finding the 10th root of 2 is a long and involved process — especially compared to dividing a number by 10. Similarly, dividing two large numbers is much harder than taking numbers away.

And that's where logarithms come in. By compiling a large table of solutions (L) to the equation $N = 10^7 (1 - 10^{-7})^L$ for any number N, which is what Napier did, it's possible to turn a difficult power or root question into a much easier multiplication or division question (using power laws), or to turn a long-winded multiplication or division problem into a much simpler addition or subtraction.

And yes, you could use it repeatedly if needed — taking logarithms twice could turn a root into a subtraction!

Napier's choice of base for logarithms is rather inconvenient — although what he worked out was correct and useful. It's really a logarithm taken away from an arbitrary constant, and the whole thing needed to be discreetly tidied up.

In the time before calculating devices, this was a major breakthrough and opened up the way for the slide rule (invented by William Oughtred in 1622) — and an entirely new constant.

Tables of logarithms were used to help make calculations go more smoothly.

LEONARD EULER

There's a joke in math that everything is named after the second person to discover it, or else nearly everything would be named after Leonard Euler.

The work of Leonard Euler (1707–1783) included introducing familiar forms of notation.

Even so, the Wikipedia page "List of topics named after Leonard Euler" has somewhere around a hundred entries. He was legendarily prolific. With the help of scribes, he produced over 50 papers in 1775 alone. When I was a researcher, the benchmark for competent productivity was maybe three or four per year. Euler's collected mathematical works span nearly 80 volumes.

Tutored as a child in Basel by Johann Bernoulli, one of Europe's greatest mathematicians, Bernoulli encouraged Euler's father to drop ideas of priesthood for his son in favor of math.

Aged about 20, he entered the Paris Academy Prize Problem for the optimal placement of masts on a ship. He didn't *check* that his arrangement was the best, he proved it, which he considered enough. He won second prize.

Ignoring (for the moment) his work in calculus, geometry, algebra, trigonometry, number theory, physics and topology (which he invented), it's almost impossible to do serious math without using some kind of notation that Euler introduced, whether it's $f(x)$ for a function, sin, cos or tan of an angle; the Greek letter Σ for a summation; or i for the square root of -1.

He also popularized π, although he wasn't the first to use it (that honor goes to the Welshman William Jones in 1706) — and, most famously, *e*.

e is an irrational number, somewhere in the region of 2.718281828459045. It can also be defined by an infinite sum:

$$\frac{1}{0!} + \frac{1}{1!} + \frac{1}{2!} + \frac{1}{3!} + \ldots = 1 + 1 = \frac{1}{2} + \frac{1}{6} + \ldots$$

It also has the lovely property that the gradient of the curve $y = e^x$ at any point is equal to the value of y at that point.

It crops up again and again in nature, science and economics, to the point that it's the most obvious choice of base for logarithms in general. Logarithms base *e* are known as *natural logarithms*.

A catalogue of Euler's work would be longer than this book, but I'll point out a couple more digressions: the invention of topology (top- means "pinnacle," -logy means "study," and the -o- means nothing) and the relationship between the number of vertices, edges, faces and volumes in a complicated shape.

Curves occur constantly in nature where, even at microscopic level, it is impossible to find a straight line.

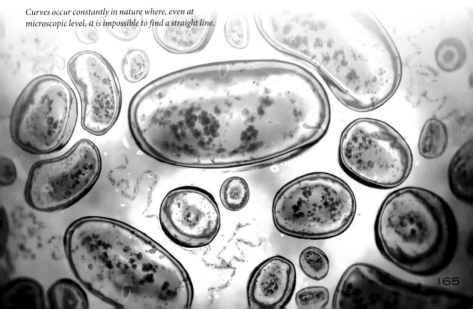

THE BRIDGES OF KÖNIGSBERG

The Prussian city of Königsberg — which is the modern-day Russian exclave of Kaliningrad — is built across the river Pregel.

In Euler's time, seven bridges linked the two banks of the river and the two islands in it. (From here on, the banks will be considered islands for simplicity.) The residents of the city, according to the story, would amuse themselves by attempting to walk across each of the bridges exactly once.

At least, until Leonard Euler showed up and spoiled their fun. He drew a picture in which dots represented the islands, and arcs represented the bridges. He showed that, no matter where you started, the complete tour was impossible. Ignoring the start and end points, any time you land on an island, you need to leave it again by a different bridge, so the intermediate bridges would need to have an even number of bridges attached to them — and in general, an *Eulerian path* exists only if there are at most two "odd nodes." Königsberg had four odd nodes, and so no such tour existed.

This abstraction of a problem of geography to one of a simple picture was a revolution and gave birth to the fields of *graph theory* and *topology* — the study of the essence of shapes.

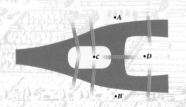

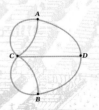

Euler looked at the map of Königsberg as a diagram with the river banks at A and B, the island at C and the peninsula at D. It was clearly impossible to visit each point crossing each bridge only once.

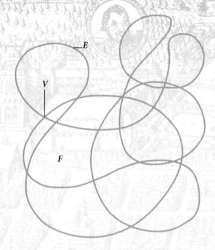

This is true for any well-behaved two-dimensional doodle, and can be extended into three dimensions, both in terms of a shape's faces and in terms of the number of separate volumes contained inside a 3D doodle involving surfaces in place of lines.

These *Euler characteristics* are a fine example of topological invariants: they're true, no matter what the shape is — topology loves to make statements about shapes in general, rather than specific shapes.

EULER'S GEM

Draw a doodle. Go ahead, I'll wait here — nice, smooth curly curves. Finished?

OK, now count the number of places the curve crosses itself, call it *V*. Now, count the number of *arcs* — the number of curve sections you have (here, that means any bit of curve between two crossing points. I find it helps to mark the ones you've counted.) Call that number *E*. Now count the number of spaces inside your curve, call it *F*.

Actually, don't bother: it's $E - V + 1$. (Purists would also include the outside of the curve as a space, in which case replace the 1 with a 2).

Name	Image	Vertices V	Edges E	Faces F	EC: V-E+F
Tetrahedron		4	6	4	2
Hexahedron		8	12	6	2
Octahedron		6	12	8	2
Dodecahedron		20	12	12	2
Icosahedron		12	30	20	2

BENFORD'S LAW

Give someone with too much time on their hands a list of 100 towns and cities, ranging from the enormous to the tiny, and have them estimate the population of each.

It wouldn't be much of a trick to compare the guess list with a real list and tell which was which, even without knowing the towns on the list, so I'll make the trick harder — just give me the first digit of each guess, and the first digit of the real populations, and shuffle each list up as much as you like. I'll still be able to tell you which list is which.

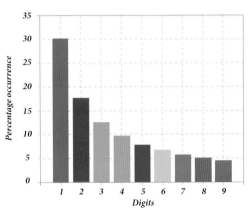

The graph shows the percentage of numbers that start with each digit, according to Benford's Law.

The thing is, humans are pretty bad at generating random numbers, or even generating numbers with a pattern to them. We tend to avoid patterns. It's easy to tell the difference between a fake set of coin tosses and a real one, as we instinctively avoid runs of more than two or three heads or tails in a row. One of the patterns, Benford's Law, is especially hard to fake.

In all likelihood, your fake list of populations has about 10 numbers that start with 1, 10 that start with 2, and so on up to 10 that start with 9. It'll vary slightly, of course. Meanwhile, the real one will have more like 30 numbers that start with 1, and five that start with 9.

It's not restricted to populations — you could do the same with a hundred rivers (it doesn't matter if you measure them in miles, kilometers, inches or anything else), a hundred entries on a financial balance sheet or any list of numbers that spans several orders of magnitude and doesn't have any particular biases in it.

Around 30 percent of the entries will start with a 1, about 18 percent with a 2, down to 4.6 percent starting with a 9.

The proportion of numbers starting with any digit n is:

$$\log_{10}((n+1)/n)$$

The reason behind this isn't entirely clear, but if you think about doubling numbers, anything that starts with a 5, 6, 7, 8 or 9 will begin with a 1 when you double it, so you would expect those numbers to be less common than 1 as lead digits.

To get nine at the start of a number by doubling, you must start with something between 4 and 5 (and in the upper half of that range), so there are fewer numbers "feeding in" to 9 than into 1.

The law is especially handy for spotting when an election has been rigged or a country's finances have been manipulated, although Frank Benford wasn't the first to discover it when he published a paper about it in the 1930s.

Simon Newcomb noticed that his table of logarithms was much more heavily used in the "1" section than in the "9" and wondered why that was — before proposing exactly the law that later became known as Benford's.

CHAPTER 8
THE CURIOUS MATH OF
ALICE IN WONDERLAND

An equation is graffitied onto a Dublin bridge, the shape of the universe is fundamentally altered, and a little girl falls down a rabbit hole.

In Lewis Carroll's Alice's Adventures in Wonderland, *Alice falls into a world where the normal rules of math seem not to apply.*

THE DISCOVERY OF THE QUATERNIONS

On October 16, 1843, William Rowan Hamilton and his wife were walking along the Royal Canal in Dublin, when he suddenly stopped dead in his tracks.

In an ideal world, he would have shouted "Eureka!" and jumped in, but instead he did what any sensible person would do: he took out a penknife and carved an equation into the nearest bridge — Brougham Bridge, now called Broom Bridge.

Hamilton, the Royal Astronomer of Ireland, had been looking for a way to extend complex numbers into three dimensions, without a great deal of progress. Have you ever noticed that your best ideas come when you're in the middle of something completely unrelated?

This was precisely what had happened to Hamilton. His breakthrough was to realize that what he had in mind couldn't possibly work in three dimensions. Four, on the other hand …

The equation he carved into the bridge was:

$$i^2 = j^2 = k^2 = ijk = -1$$

Hamilton used three imaginary numbers (i, j and k), which he called *quaternions*, to represent the three dimensions of space, and a real number to represent time. It wasn't enough just to go beyond three dimensions, Hamilton also had to abandon *commutativity*.

When you're learning your times tables, you can save yourself a lot of effort by knowing that 7×4 is the same thing as 4×7, for example. With real numbers, it doesn't matter which way round you multiply two numbers. The same goes for adding: $x + y$ and $y + x$ are always the same. This isn't true

Here as he walked by
on the 16th of October 1843
Sir William Rowan Hamilton
in a flash of genius discovered
the fundamental formula for
quaternion multiplication
$$i^2 = j^2 = k^2 = ijk = -1$$
& cut it on a stone of this bridge

A plaque on Broom Bridge in Dublin, commemorates Hamilton's moment of inspiration, when he carved his formula into the bridge.

for multiplying Hamilton's quaternions, though — whereas $ij = k, ji = -k$. Changing the order of the multiplication of two quaternions changes the sign of the result.

Hamilton solved his problem by abandoning two assumptions he (and everyone else) had made about the

problem — so I'd be inclined to forgive his act of vandalism.

Taoiseach Éamon de Valera was also so inclined. He unveiled a plaque on Broom Bridge in 1958 to commemorate Hamilton's moment of triumph.

However, not everyone saw him as a mathematical hero.

QUATERNION APPLICATIONS

The quaternions i, j *and* k *are the idea behind the* **i**, **j**, **k** *notation for vectors — you can represent any point in our standard three-dimensional space using three coordinates, usually written* (x, y, z).

In vector notation, though, you can write the same thing as $x\mathbf{i} + y\mathbf{j} + z\mathbf{k}$. Using quaternion multiplication on two vectors has some extremely useful applications.

If you simply multiply two vectors together as if they were quaternions, you get another quaternion, usually involving a real number as well. Hamilton, for no good reason, called this "time".

For example, if you worked out $(2i + 3j + 4k)(3i - 2j - 4k)$, you would get $16 - 4i + 20j - 13k$. The "time", in this case 16, is the negative of the scalar product of the vectors, a measure combining the lengths of the vectors and the angle between them. The vector part, $-4i + 20j - 13k$, is the vector product, which combines the lengths of the

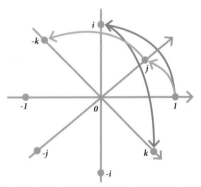

The product of quaternion units can be represented graphically as a 90° rotation in 4D-space.

vectors, the angle between them and — very handily — a vector at right angles to both of the original vectors.

Many relationships in physics can be written using the scalar or vector product, and having an algebraic way to compute them is especially useful.

Their main benefit, though, is that you can use them to rotate a vector around an arbitrary axis. Given a point in space (a, b, c), an axis to rotate around (x, y, z) and an angle t to spin, you can calculate where your point ends up by working out the quaternion product of:

$$(\cos(t) + \sin(t)[x\mathbf{i} + y\mathbf{j} + z\mathbf{k}]) \cdot (a\mathbf{i} + b\mathbf{j} + c\mathbf{k}) \cdot (\cos(t) - \sin(t)[x\mathbf{i} + y\mathbf{j} + z\mathbf{k}])$$

That looks a mess — but it's significantly easier than the traditional method of repeatedly rotating the axis to somewhere convenient, doing the desired rotation and rotating back.

Not only is this kind of rotation easier to work out, it's much easier to work with if you want a smooth rotation (for example, if you're animating a computer game) and avoids a technical

In Lewis Carroll's Through the Looking-Glass, *Alice finds an inverted book in which words are written in mirror-writing. She holds a mirror to the page to reveal the nonsense poem "Jabberwocky." This action would be represented mathematically as reflection across an axis.*

problem called gimbal lock. This, sadly, is entirely unrelated to the gimbling that the slithy toves do in Lewis Carroll's "Jabberwocky" poem.

*German mathematician
Carl Friedrich Gauss*

NON-EUCLIDEAN
GEOMETRY

For somewhere around 2,500 years, Euclid's Elements *was pretty much unquestioned
as the source of all geometrical knowledge.*

To question Euclid was tantamount to blasphemy — apart from one question: whether you could prove the parallel postulate from the other, more elegant assumptions he made. The parallel postulate says:

> *If a line segment intersects two
> straight lines forming two interior
> angles on the same side that sum to*

> *less than two right angles, then the
> two lines, if extended indefinitely,
> meet on that side on which the angles
> sum to less than two right angles.*

Punchy, isn't it? Euclid, apparently, spent time trying to prove it, and so did two-and-a-half millennia of other mathematicians, without success. In the early 19th century, several people had the same idea: what if, instead of trying to prove it directly, you assume the opposite and see what happens? If you end up with something impossible, then you have a contradiction, which means the assumption must have been wrong.

János Bolyai was one of them, finding it possible to replace the parallel postulate with something else, and still get geometry that … well, "made sense" would be a bit strong.

If you did one thing, you got a *projective* geometry where every pair of lines would eventually meet. If you made another assumption, you got a *hyperbolic* geometry where several different lines through a given point would never meet. Bolyai's work was absolutely revolutionary.

Bolyai's father, who was also an accomplished mathematician in his own right, proudly wrote to Carl Friedrich Gauss, who wrote straight back to say something along the lines of, "This stuff is brilliant — but I would say that, I thought of it first. I've been thinking about it for years, actually."

To be fair, Gauss had in fact thought of it first. He just hadn't published.

Discouraged, Bolyai left his work unpublished for another decade. Meanwhile, Nikolai Ivanovich Lobachevski figured out largely the same thing and, as Tom Lehrer's song puts it:

> *In Dnepropetrovsk, my name*
> *is cursed, when he finds out*
> *I published first!*

Not just in Dnepropetrovsk was Lobachevski's name cursed: in Oxford, there was a tutor who didn't think much of this new-fangled geometry.

Lewis Carroll did not like hyperbolic geometry at all.

NON-EUCLIDEAN GEOMETRY: APPLICATIONS

If there's one thing mathematics is really good at, it's coming up with solutions in search of a problem.

Pure mathematicians, an odd bunch, are fiercely proud to be working on topics without an obvious application, and many hope that their work will remain forever perfectly pure and abstract. Bless their hearts.

Non-Euclidean geometry is a perfect example of a "useless" discovery which turned out to be extremely important (as are, for example, elliptic curves).

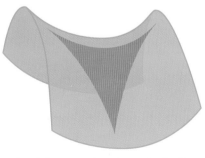

In hyperbolic geometry, the angles in a triangle add up to less than 180°.

It turns out that science was missing at least two examples of non-Euclidean geometry, one of which was right under its nose.

The surface of the Earth itself can be described using projective geometry. If you define a point as any two locations diametrically opposite each other, and a line as the great circle between any two points, you get a perfectly sensible way of doing geometry on a sphere — only there are now triangles with more than 180° in them.

Consider the triangle starting at the North Pole, traveling due south along the 0° meridian through London until it reaches the Equator in the Gulf of Guinea, then east until it reaches a point 90°E somewhere in the Indian Ocean, then returns to the pole — that's a triangle with three right angles!

Triangles on the surface of the Earth have internal angles adding up to more than 180°.

The other example was not quite so close to home or obvious: whereas Euclidean geometry works pretty well on a small scale, Einstein showed that, on astronomical scales, space–time exhibits a weird curving behavior around massive objects — meaning that Euclidean geometry doesn't hold true. In fact, the best model we have says it's a pseudo-Riemannian manifold, but I don't know what that means and neither should you.

POINCARÉ DISK MODEL

A lovely way to visualize hyperbolic geometry is the Poincaré Disk Model. A point is, unsurprisingly, represented as a point; a line, on the other hand, is represented as an arc of a circle.

On a Poincaré hyperbolic disk, a line is an arc of a circle whose ends are at right angles to the edge of the disk.

Triangles in hyperbolic space have angles that sum to less than 180°. Several lines can pass through a particular point that are all parallel to another line, in the sense that they do not intersect (in Euclidean geometry, where the Parallel Postulate holds, exactly one parallel line should exist).

Dutch artist M.C. Escher made great use of this model with his woodcuts in the Circle Limit series. Fascinated by tessellations after a visit to the Alhambra Palace in 1936, Escher wrote to Canadian mathematician H. S. M. Coxeter about his work on representing the hyperbolic plane.

Circle Limit 1 shows stylized fish arranged on the disk, while *Circle Limit II* shows a series of crosses. At this point, Escher appears to have realized he could do much more and *Circle Limit III* has multicolored fish swimming across the plane.

Personally, my favorite is the last in the series, *Circle Limit IV*, which shows angels and demons arranged alternately on the disk.

Escher was inspired by the geometric decorations at the Alhambra Palace in Granada, Spain.

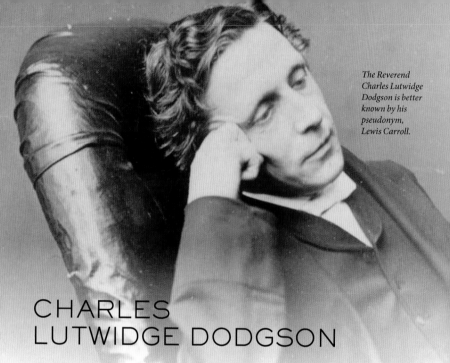

The Reverend Charles Lutwidge Dodgson is better known by his pseudonym, Lewis Carroll.

CHARLES LUTWIDGE DODGSON

Charles Lutwidge Dodgson (1832–1898) was a Mathematical Lecturer at Christ Church College, Oxford, and specialized in geometry, logic, linear algebra and recreational mathematics.

If you wanted a caricature of a math professor, you wouldn't go far wrong in picking Dodgson. Plagued by waking up in the middle of the night with a brilliant idea, having to make the hard choice between getting up to light a candle and going back to sleep at the risk of losing his idea, he invented the *nyctograph*, a template for taking notes in the dark.

Disappointed in the first-past-the-post system, he came up with a fairer way to decide elections. He came up with a better way than straight knock-out to arrange a tennis tournament. He was a pioneer of photography. He invented codes. He invented games and puzzles. He's even got a pretty good claim to have invented double-sided sticky tape.

Remarkably, though, he's not best-known for his mathematical or inventing endeavors. He's better known for his literary work, published under a pen name derived from his forenames: Lutwidge is a corruption of Lewis, and an alternate form of Charles is Carroll.

As Lewis Carroll, he wrote some of the most famous children's stories of his (or any) time. Phrases and characters from *Alice's Adventures in Wonderland* and *Through The Looking Glass* have become common parts of the English language — from "falling down a rabbit-hole" to mean getting caught up in nonsense, to "portmanteau words" that combine two separate words into one.

Alice in Wonderland isn't just a fantastically bizarre children's story, it's a mathematical protest novel. If there were two things Dodgson couldn't stand, they were non-Euclidean geometry and quaternions — and he sends Alice on a quest to show how absurd they are.

Apart from the *Alice* books, Carroll wrote several long poems (*The Hunting of the Snark* and *Phantasmagoria*, among others), dialogues (including *Euclid and his Modern Rivals* and *What the Tortoise Said to Achilles*) and children's stories (*Sylvie and Bruno*), although never quite repeating his success with *Alice*.

Dodgson died of pneumonia aged 65 and is buried in Guildford, England.

Among his diverse areas of interest, Dodgson was an early pioneer in photography and he took many self-portraits such as this one when he was aged 23.

THE NYCTOGRAPH

One of the legends of the St Andrews math department is that, when he was a PhD student, Professor Alan Hood sat bolt upright in bed one night, startling his wife. He said "Of course! That's why it's 2π!" and fell straight back to sleep.

Reminded by Mrs Hood of this in the morning, he shook his head. Naturally, he had no recollection of either the incident, the explanation that had come to him in his sleep, or even what was supposed to be 2π.

Had the Hoods' bedroom been equipped with a nyctograph, perhaps Professor Hood would have yet another scientific discovery to his name.

It's a sheet of cardboard with 6mm squares cut into it, in two rows of eight. You use it as a template for jotting down notes without having to go to the bother of turning the light on — which was, in Dodgson's day, a big deal, especially if you were going to turn it straight off again (they didn't have telephones in those days, let alone mobiles with note-taking apps!).

Dodgson came up with a shorthand alphabet written using lines around the edges of the squares, and dots in the corners. He was rather proud that 23 of the 26 letters looked roughly like the

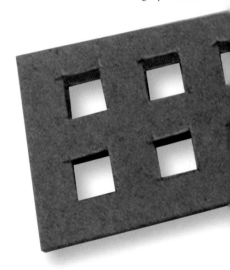

ones they were supposed to represent. All of the letters have a large dot in the top-left corner, whereas punctuation marks have them in the lower right.

Alan Tannenbaum, a member of the Lewis Carroll Society of North America, has released a version of *Alice in Wonderland* using Dodgson's alphabet — although, as he points out, you'll probably need to turn on a light to read it.

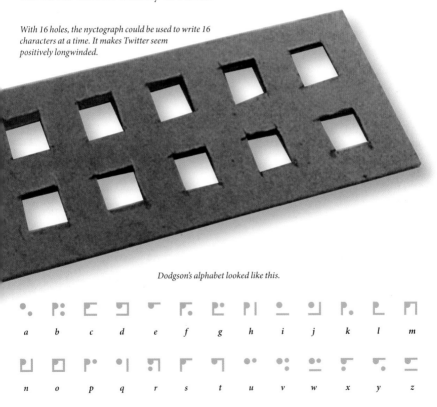

With 16 holes, the nyctograph could be used to write 16 characters at a time. It makes Twitter seem positively longwinded.

Dodgson's alphabet looked like this.

a	b	c	d	e	f	g	h	i	j	k	l	m

n	o	p	q	r	s	t	u	v	w	x	y	z

DODGSON'S ELECTIONS

Dodgson's method for deciding elections is much easier to explain than it is to work out in practice.

The ideal situation in an election is for there to be one candidate who would win head-to-head against each of the other candidates. If there's such a candidate — known as a *"Condorcet candidate"* — he or she ought to be the winner in a proper election. The first-past-the-post method (FPTP) used for parliamentary elections in Great Britain, the U.S. and elsewhere, doesn't guarantee that a Condorcet candidate wins.

There's not always a Condorcet candidate. Imagine an election between the Rock Party, the Scissors Party and the Paper Party. The Rock Party has 20 voters who prefer Rock to Scissors but prefer Scissors to Paper. The Scissors Party has 25 voters who prefer Scissors to Paper and Paper to Rock. The Paper Party has 30 voters who prefer Paper to Rock and Rock to Scissors.

Overall, there are 50 voters who prefer Rock to Scissors against 25 who don't,

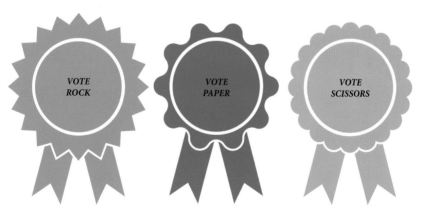

but Paper would beat Rock 55–20. Similarly, each of the other parties would beat one of the others but lose to the third. Thus, there is no Condorcet candidate. Using Dodgson's method, the winning candidate is the one who'd need the fewest people to change their votes in order to become a Condorcet candidate.

In this case, I've picked the numbers to make it easy for me to work it out — if 18 Scissors voters put Rock ahead of Paper, Rock would be the Condorcet candidate. If 13 Paper voters put Scissors ahead of Rock, Scissors would win.

But only 8 Rock voters would need to put Paper ahead of Scissors for Paper to be the Condorcet candidate, so Paper would win, both by FPTP and by Dodgson's method. In general, working out exactly how many votes need to change for a candidate to win is an *extremely* hard problem.

According to a paper by Bartholdi, Tovey and Trick, it's NP-hard — which means that as the number of candidates grows, the number of calculations grows extremely quickly, and for even a modest by-election field of 20 candidates, the best computers would take many years to solve it.

As the authors nicely put it, it's likely that by the time a winner had been determined by Dodgson's method, it would be time for a new election.

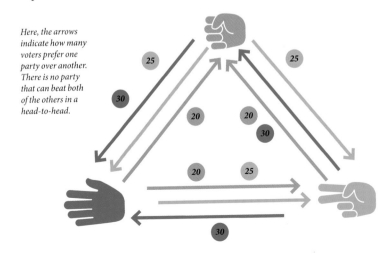

Here, the arrows indicate how many voters prefer one party over another. There is no party that can beat both of the others in a head-to-head.

ALICE AS A PASSIVE-AGGRESSIVE PROTEST NOVEL

Depending on your point of view, projective geometry (one of the non-Euclidean geometries proposed by Bolyai) has an enormous bug or an interesting feature.

In the regular Euclidean geometry Dodgson grew up with, if you move an object around, it stays the same size.

That, to most people (Dodgson included) is as it should be. In projective geometry, moving an object around changes its apparent size. This is, well, it's downright weird, especially if Euclidean geometry is all you've ever known.

And it's a real threat, if you're a mathematical lecturer and Euclidean geometry is all you've ever taught. Suddenly, everything you thought had a solid foundation has been overturned.

You'd probably argue strongly against them, as Dodgson did, in the journals of the day and you'd be out-argued at every turn by people who are following completely different rules — but which seem to make sense to them.

You might call it an *Alice in Wonderland* situation — because that's exactly what

Alice changes size as she moves around, as objects do in projective geometry.

Perspective lines are a form of projective geometry. It makes us think that we are looking at a 3D space. Objects higher up on the image are perceived to be bigger as we assume they are further away. These two figures are in fact the same size.

happens in *Alice*. Alice is continually changing size, and continually trying to reason with infuriating creatures who refuse to accept her point of view about reality.

So far, so circumstantial, but stick with me — it's the quaternions that are the real clincher.

Dodgson, being a linear algebra man, knew an awful lot about *matrices* — arrays of numbers that can be used to represent points and transformations in as many dimensions as you like. Dodgson condensation, a method for working out a matrix's determinant, is yet another of Charles Lutwidge Dodgson's inventions. One of the things you can do with matrices is rotate a point around an arbitrary axis — but it's fiddly! You usually need to perform five or even seven matrix operations, each with 27 separate calculations.

Quaternions, meanwhile, generally needed something around 30. The pesky quaternions posed another real threat to Dodgson's teaching.

Remember Hamilton's problem — that if you only had three imaginary numbers, you ended up going around in circles? Only by adding "time" could he make his formulation work properly.

At the Mad Hatter's tea party, there are *meant* to be four participants: the Mad Hatter, the March Hare and the Dormouse have all made it just fine. However, the final guest, Time, has had a big falling out with the host, and hasn't turned up. The remaining partygoers move in circles around the table to get to the clean crockery.

Remember also that quaternions don't *commute* — *ij* and *ji* give different answers. One of my favorite parts of the Tea Party goes like this:

> "Then you should say what you mean," the March Hare went on.
> "I do," Alice hastily replied; "at least — at least I mean what I say — that's the same thing, you know."
> "Not the same thing a bit!" said the Hatter. "You might just as well say that 'I see what I eat' is the same thing as 'I eat what I see!' "
> "You might just as well say," added the March Hare, "that 'I like what I get' is the same thing as 'I get what I like!' "
> "You might just as well say," added the Dormouse, who seemed to be talking in his sleep, "that 'I breathe when I sleep' is the same thing as 'I sleep when I breathe!' "

The Mad Hatter represents Carroll's distaste for quaternions.

In Alice's world, things commute. At the Mad Hatter's Tea Party, like with quaternions, the order you put things in matters a great deal.

EUCLID AND HIS MODERN RIVALS

For some reason, trying to make his point that Euclid was the be-all and end-all of geometry teaching in the form of a children's book somehow didn't put an end to people studying non-Euclidean geometry and even — the horror — using different textbooks to teach Euclidean geometry.

In 1878, Dodgson published *Euclid and His Modern Rivals*, clearly setting out in dialogue form his opinion of some 13 alternative geometry texts that had crossed his desk.

Time is missing from the Mad Hatter's tea party, causing the guests to move in circles.

Dodgson's book (he calls it "little" in the introduction, although it runs to some 300 pages of fairly dense dialogue) unsurprisingly comes down on the side of Euclid.

Given that one of the characters is the ghost of Euclid, most mathematicians would tut and make nasty remarks about assuming your result. The whole of Act II is dedicated to the defence of the parallel postulate, which gives an idea of how strongly he felt about it.

Euclid and His Modern Rivals, even for a book by Lewis Carroll (in his day job), would probably be too obscure for me to mention but for one thing: it does have a place in popular culture. When Wikipedia was first set up in 2001, a passage from it was put under the magnifying glass in the online encyclopedia's logo.

The first Wikipedia logo played with Carroll's attachment to Euclid.

A TANGLED KNOT

One of my prized possessions is a *Complete Works of Lewis Carroll* given to me by my parents as a schoolchild.

Once I'd devoured the *Alice* stories and given up on *Sylvie and Bruno*, I started reading *A Tangled Tale*. It was a revelation: silly stories about math! I couldn't do the puzzles in the stories, which Carroll called Knots (in fact, I had trouble understanding them), but I appreciated the general idea.

There were stories of baffled knights, belligerent aunts, bizarre family trees, in which Carroll concealed ("like medicine … in the jam of our childhood") ingenious puzzles, inviting solutions from readers of the *Monthly Packet* magazine they were originally published in.

Carroll would read the solutions to his Knots sent in by his readers and rate each of them.

He'd then analyze the solutions received and classify them according to how fully they'd answered.

The last Knot of the 10-part series, *The Change of Day*, is never answered. Carroll himself claimed to have been puzzled by it.

CHAPTER 9
THE INFINITE, THE UNDECIDABLE AND THE COMPUTER

It is discovered that there are infinitely many kinds of infinity, the whole mathematical pyramid comes tumbling down, and out of the destruction comes a machine that can do almost anything.

There is an infinite variety of infinities.

GEORG CANTOR

A mother asks her son, "What's the biggest number?" The son thinks for a moment, and says, "A trillion!"
"But what about a trillion and one, dear?" The son looks crestfallen for a moment, but then perks up — "Well, I was close!"

There are, of course, infinitely many numbers — if ever you think you've got to the end of the numbers, you can try adding one more; there will always be room for it. There are also infinitely many numbers between 0 and 1 — there are a whole load of fractions, for example, like ½ and ¾ and $\frac{9}{17}$, as well as irrational numbers such as $^1/\pi$ and $\sqrt{(1/2)}$ and $\frac{3}{e}$, and Champernowne's number, **0.12345678910111213141 5...** There are others.

The German mathematician Georg Cantor (1845–1918), in the late 19th century,

Georg Cantor asked awkward questions with surprising answers.

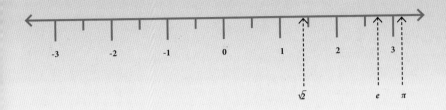

A real number line with some old friends also marked on — but can there really be as many points between 0 and 1 as there are numbers you can count?

managed to use mathematical reasoning to provoke a surprisingly violent reaction. He asked, "Are there as many counting numbers as there are points between 0 and 1?" and found — completely counterintuitively — that there are different sizes of infinity. In some sense, there are more real numbers between 0 and 1 than there are whole numbers starting from 1 and going up without limit.

A surprisingly violent reaction? Well, yes. Henri Poincaré, one of the world's greatest mathematicians, called Cantor's work a "grave disease" infecting mathematics, while Cantor's old boss Kronecker didn't exactly mince his words. According to him, Cantor was a "corruptor of youth," and a "scientific charlatan" — and, being a pretty powerful man, effectively blacklisted Cantor from working at Berlin, one of the top universities in Germany (and hence in the world) for math at the time. Ludwig Wittgenstein said Cantor's work was "laughable" and "utter nonsense," although he's not recorded as having explained why it was wrong.

In mathematics the art of asking questions is more valuable than solving problems.
Georg Ferdinand Ludwig Philipp Cantor

Perhaps as a result of this vitriolic abuse, Cantor suffered from severe depression and was admitted to a sanatorium in 1884. There have even been suggestions that he may have had undiagnosed bipolar disorder.

He recovered enough to continue his research, but never managed to match the standard of his earlier work.

His final two decades are a tale of woe. His son, Rudolph, died suddenly; he felt humiliated at conferences when others undermined his work; and he traveled to a conference at St Andrews, Scotland, in the hope of meeting Bertrand Russell, but Russell didn't attend. Then, during World War I, he suffered from malnourishment and — adding insult to injury — the cancellation of a public celebration of his 70th birthday. He died in a sanatorium in 1918.

THE FLAVORS OF INFINITY

How can you tell whether two infinities are different? To answer that, you need to be able to tell when sets of objects are the same size. Obviously, you could count the number of elements in each set and see if the numbers match.

Good answer, that — as long as you're dealing with finite sets. What if your sets have infinitely many elements, like the whole numbers, or the fractions, or the real numbers? Cantor was the first to come up with a reasonable answer: two sets have the same size (cardinality) if you can arrange them so that each element in one set corresponds to exactly one element in the other — an idea called one-to-one correspondence.

Finite sets, like football teams, are easy to count.

Plagued by ill-health, Cantor regularly sought refuge in the tranquillity of the Harz Mountains.

Think about the even numbers: 2, 4, 6, 8, 10 … Although it might *seem* that there are half as many of them as there are whole numbers — after all, whole numbers are alternately even and odd, so you'd think that half of them are even and half odd — in fact, there are just as many even numbers as there are whole numbers: you can associate each even number (an element in the first set) with the unique whole number with half its value (an element in the second set) — and every element in each set is matched off one-to-one with an element in the other. They have the same cardinality.

Even more oddly, the set of fractions has the same cardinality as the set of the whole numbers. You can arrange all of the possible fractions in a sort of order: all of the fractions whose tops and bottoms add up to one (only $\frac{0}{1}$); followed by all of the fractions whose tops and bottoms add up to two ($\frac{1}{1}$ and $\frac{0}{2}$); then the ones that add up to three ($\frac{2}{1}$, $\frac{1}{2}$ and $\frac{0}{3}$), and so on — eventually, every fraction will appear in the list. Even allowing for duplicates ($\frac{0}{1}$, $\frac{0}{2}$ and $\frac{0}{3}$ are

the same number, 0), it's possible to create a (complicated) one-to-one correspondence between the whole numbers and the fractions.

$$1 \longleftrightarrow 2$$

$$2 \longleftrightarrow 4$$

$$3 \longleftrightarrow 6$$

$$4 \longleftrightarrow 8$$

$$5 \longleftrightarrow 10$$

$$6 \longleftrightarrow 12$$

$$7 \longleftrightarrow 14$$

$$8 \longleftrightarrow 16$$

$$9 \longleftrightarrow 18$$

$$10 \longleftrightarrow 20$$

Pairing up whole numbers with even numbers gives a one-to-one correspondence.

Some of the notation of modern set theory includes:

$\mathbb{N}$ = the set of natural numbers

$\mathbb{Q}$ = the set of rational numbers

$\mathbb{R}$ = the set of real numbers

$\mathbb{Z}$ = the set of integers

- The symbol for showing that one set is a subset of another set is $\subset$, so that:
$N \subset Z \subset Q \subset R$

- Curly Brackets {} are used to show a collection of elements in a set:
$A = \{3, 7, 9, 14\}$

- The number of members in a set is shown as $||$ or #:
If $A = \{3, 7, 9, 14\}$, then $|A| = 4$ or $\#A = 4$

- The cardinality of infinite sets is denoted by $\aleph$:
$|N| = \aleph_0 \qquad |R| = 2^{\aleph_0}$

THE HILBERT GRAND HOTEL

German mathematician David Hilbert took up Cantor's infinity ideas, demonstrating them with an illustration involving his Hilbert Grand Hotel. Imagine a hotel with an infinite number of rooms — just your regular, standard, countable infinite number of rooms, starting at 1 and never ending; they're identical, but for the number on the door (we shall gloss over what happens when the number is too big for the door). It's a busy night at the Hilbert Grand, and all of the rooms are occupied.

The bell rings at reception, and a tired-looking traveler stands there.

> "Do you have a room for the night?" she asks.

The receptionist puts on his most apologetic face:

> "I'm terribly sorry, madam, all of our infinitely many rooms are occupied."

The traveler smiles a weary smile.

> "Ah, but that's why I came here! I'm a mathematician, so I knew you would be able to find a room for me. All you need to do is to move the occupants of room 1 into room 2. The people in room 2 can move into room 3, and in general room n moves to room n+1."

> "I see," says the receptionist.

> "Everyone simply moves into the next room, so you can take room 1, and everyone else still has a room. An excellent solution, I'm certain none of our infinitely many guests will mind moving rooms at this hour."

The receptionist arranges this, and sits back down with his copy of *The Mathematics Bible*, to entertain him for the rest of his night shift. A moment later, the bell rings again.

> "How can I help you, sir?"

"Hello, there. I'm a tour guide and I've got a bus with infinitely many people on board, all of whom want a room for the night. Can you help?"

"I'm terribly sorry, sir, all of our infinitely many rooms are currently occupied… oh, let me guess, you're a mathematician, too."

"Correct," says the tour guide, twirling his garish umbrella. "There's a simple solution. Move the occupant of room 1 into room 2. The occupant of room 2 can move to room 4; room 3 goes to room 6; and in general, room n goes to room 2n."

"So now all of the even-numbered rooms will have someone in them, but the odd-numbered rooms will be free," nods the receptionist.

"Correct again! And, since there are infinitely many of those, everyone in my tour group will have a room."

"A marvellous solution, sir. I'm certain that none of our guests will mind being moved again."

At Hilbert's Grand Hotel, the infinitely many rooms may all be full, but the hotel can still accommodate infinitely many extra guests, simply by shifting each existing guest from room n into room 2n.

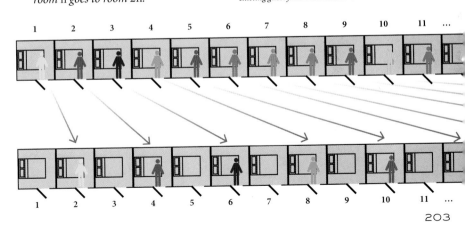

The mathematical symbol for infinity finds a three-dimensional analogy in a Möbius strip — a surface that has only one side. If you were to walk along a Möbius surface, you would traverse the whole surface before returning to your starting point.

Where Cantor's work became controversial was when he started to think about the real numbers, which he proved *couldn't* have a one-to-one correspondence with the whole numbers.

His *diagonalization argument* showed that any scheme you tried would inevitably fail. There was a difference between the cardinality of the whole numbers (*countably infinite*) and the cardinality of the reals (*uncountably infinite*). Cantor also showed that there were infinitely many possible varieties of infinity, and he wondered whether there were any infinities between the size of the whole numbers and the size of the reals. The solution to that problem would take 50 years, and has the very odd answer of "it can't be proved either way."

CANTOR'S DIAGONALIZATION ARGUMENT

To prove that there are more real numbers than whole numbers, Cantor first assumed that the whole numbers and the real numbers could be lined up with one another, as we saw could be done with whole numbers and even numbers.

So, in a certain list, each real number between 0 and 1 could be paired one-to-one with a whole number. For instance:

Each real number's decimal expansion can potentially go on forever, hence the "…"

Next, you make a real number from the digits bolded in the list: the first digit in the first row, the second digit in the second row etc, so here we have:

0.25625 …

Now, it is possible to construct another real number that differs from this one in each corresponding place: for instance, by changing the digit 5 to 1, and changing all other digits to 5. This gives:

0.51551 …

$1 \leftrightarrow 0.2\mathbf{4}356\ldots$

$2 \leftrightarrow 0.1\mathbf{5}479\ldots$

$3 \leftrightarrow 0.35\mathbf{6}58\ldots$

$4 \leftrightarrow 0.875\mathbf{2}4\ldots$

$5 \leftrightarrow 0.7846\mathbf{5}\ldots$

But this new decimal expansion now differs from the first number on the list at the first decimal point, from the second number at the second decimal point, right up to any number n, where it will differ at the nth decimal point. So it does not appear anywhere on the list. This means that there are more real numbers than there are whole numbers, and they are not countable.

DAVID HILBERT

What is the first thing you notice when looking at a picture of David Hilbert (1862–1943)? Perhaps his neat, pointy beard; maybe his forbidding round spectacles.

Possibly it's the amazing hat. But if you asked someone on the street who the picture was of, they'd probably guess, "an evil psychologist?" followed by "Oh! A math professor."

Like Christian Goldbach, Immanuel Kant and the field of topology, Hilbert was born near Königsberg (now Kaliningrad), where he studied and lectured until 1895 before moving to Göttingen — at the time, probably Europe's greatest mathematical center.

Among his students and colleagues were Hermann Minkowski (who figured out much of the math behind Einstein's theories of relativity),

David Hilbert popularized Cantor's ideas and posed 23 problems of his own.

Hermann Weyl (one of the last great mathematical universalists), Ernst Zermelo (who got set theory more or less under control) and John von Neumann.

Apart from popularizing Cantor's ideas (especially by means of the Hilbert Hotel), he is best known for his 23 problems. At the International Congress of Mathematicians in Paris, on August 8, 1900, Hilbert presented 10 unsolved problems that he considered the most important in math at the time. The list later expanded to include another 13.

Some are simply stated (2. *Prove that the axioms of arithmetic are consistent* — we'll come back to that later in this chapter), whereas others require more technical language (12. *Extend the Kronecker-Weber theorem on abelian extensions of the rational numbers to any base number field* — we'll leave that one to the experts).

Many have been resolved over the last hundred years or so. Others (such as the Riemann hypothesis) remain stubbornly untouched. Still others have led to intense research activity and the occasional new field of mathematics.

Nazis burn books during the 1933 purges in Germany.

Hilbert retired in 1930, not long before the Nazi purge of the mathematics department at Göttingen, which deprived the university of (among many others) Weyl, Emmy Noether and Edmund Landau. When asked by the Minister of Education how mathematics in Göttingen was, now it was free of Jewish influence, Hilbert reportedly replied that there wasn't really any mathematics there any more.

Good, he didn't have enough imagination to become a mathematician.

David Hilbert
(on hearing that a student had dropped out to become a poet.)

WHO SHAVES THE BARBER?

The year is 1902 and Gottlob Frege is feeling rather pleased with himself — he was putting the finishing touches to Grundgesetze der Arithmetik, *the basic laws of arithmetic.*

In this worthy tome he has managed to derive all of the laws of arithmetic from a few logical axioms. It's about to go to the printers (at his own expense).

He's disturbed by the arrival of the mail; there's a letter from Bertrand Russell. This is usually a good thing,

Friedrich Ludwig Gottlob Frege (1848–1925), German mathematician and philosopher.

but this particular letter contains a bombshell: the system contains a paradox. This is very bad news for a logical system.

There's a long-standing tradition of mathematicians testing the limits of their subject by trying to create contradictions. One of the first was Epimenides of Knossos, which is on Crete, who said "All Cretans are liars." This is a paradox: Epimenides, a Cretan, is calling himself a liar; if what he says is true, then he's a liar; if what he says is a lie, then he's telling the truth about those lying Cretans.

More succinctly, statements like "This statement is false" or "The following statement is true; the preceding statement is false" cause an immediate logical problem: if they're true, then they're false, and so on.

This is pretty much the basis of what

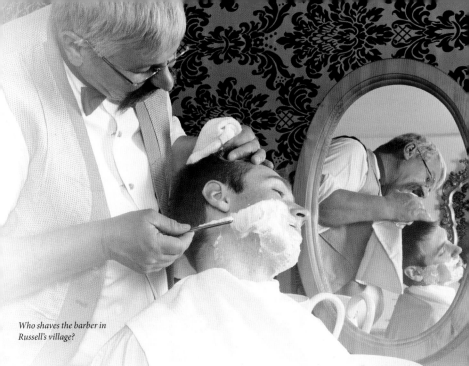

Who shaves the barber in Russell's village?

Russell had found in the Grundgesetze: he suggested "the predicate: to be a predicate which cannot be predicated of itself," and asked whether it could be predicated of itself. If it could, then it couldn't and vice-versa. A less abstract version of Russell's paradox involves a village in which the barber shaves all the men, none of the men in the village shaving themselves. The question is, who shaves the barber?

Russell wasn't the first to notice this problem with set theory — Zermelo came up with it a year before, but didn't publish. And, as we know, if you don't publish, it doesn't count.

Frege, meanwhile, hurriedly added an appendix to his book, acknowledging the paradox and attempting to fix it — but the system was fatally flawed. Russell, with his boss Alfred North Whitehead, would attempt to patch it up themselves.

Bertrand Russell and his wife, Edith,
protesting against nuclear weapons.

BERTRAND RUSSELL

When I grow up, I want to be like Bertrand Russell (1872–1970), although perhaps without the pipe-smoking.

I don't have his aristocratic background, of course, and none of my grandparents were ever Prime Minister, but it's a measure of the esteem in which I hold him that I don't begrudge him any of that.

He was largely educated by tutors and governesses, mastering several languages while young, but math was his great love. A melancholy teenager, his thoughts of suicide were tempered by his desire to do more mathematics.

In his relatively early years, he was one of the world's greatest logicians and mathematicians, trying (and failing) to build math up consistently from logical foundations in the *Principia Mathematica*.

He gained a great deal of notoriety as a political campaigner. He campaigned for votes for women and against World War I. When he refused to pay a fine for writing in support of a conscientious objector in 1916, his books were auctioned. His friends bought them back, and he was especially proud of his *King James Bible* stamped "Confiscated by Cambridge Police."

He was later jailed for campaigning against America's involvement in World War I but didn't let that stop

him from later promoting pacifism. He also campaigned against Nazism, against Stalinism, for the reform of laws against homosexuality, against the Vietnam War and, most prominently, against nuclear weapons.

Like Christabel Pankhurst (right) and Annie Kenney, Russell campaigned for women's rights.

He was also awarded a Nobel Prize for Literature, the Order of Merit (the King was reportedly rather embarrassed to be handing out medals to a jailbird), the Sylvester Medal of the Royal Society (for distinguished work on the foundations of mathematics), the de Morgan Medal of the London Mathematical Society, founded a school without any rules and was removed from a lecturing job in New York for being perceived as having a dreadful moral influence.

He was one of the rare writers who managed to be both very clever and very readable. I'd be delighted to achieve either.

I was told the Chinese said they would bury me by the Western Lake and build a shrine to my memory. I have some slight regret that this did not happen, as I might have become a god, which would have been very chic for an atheist.

Bertrand Russell studied mathematics at Trinity College, Cambridge.

PRINCIPIA MATHEMATICA

After finding the problem in Frege's work, Russell and Alfred North Whitehead set about trying to put it right, using a small set of axioms and rules to develop the whole mathematical structure — only without allowing for the construction of the paradoxical sets that had punched holes in Frege's work. The Russell and Whitehead approach to the problem was to use hierarchies of types to avoid self-referential mathematical statements (although this, too, was doomed to failure).

It was a monumental task: three volumes, published in the run-up to World War I, and it was a success in some senses. It didn't encompass all of math in the end (it got as far as the real numbers), but experts agreed the system could be used to develop further ideas — it'd just take forever, and nobody could really be bothered. Not even Russell and Whitehead, who abandoned the project after Volume III, citing intellectual exhaustion; the planned book on geometry never appeared.

To show how long it would take, I refer to the only thing that every mathematician knows about *Principia Mathematica*: on page 379 of Volume I in the first edition, the authors make the following statement: "From this proposition it will follow, when arithmetical addition has been defined, that $1+1=2$." The full proof isn't completed until page 86 of Volume II.

PRINCIPIA MATHEMATICA

TO ∗56

BY

ALFRED NORTH WHITEHEAD

AND

BERTRAND RUSSELL, F.R.S.

CAMBRIDGE
AT THE UNIVERSITY PRESS

Russell and Whitehead abandoned their great work on mathematics, realizing that it would take several lifetimes to complete!

GÖDEL DESTROYS MATH

If you have a mathematical system, you want just three simple things of it. It needs to be:

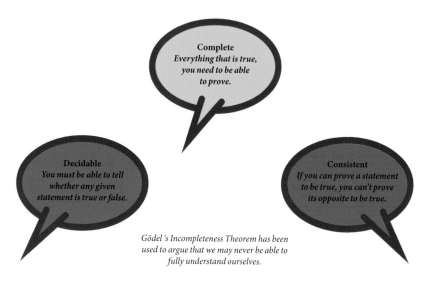

Complete
Everything that is true, you need to be able to prove.

Decidable
You must be able to tell whether any given statement is true or false.

Consistent
If you can prove a statement to be true, you can't prove its opposite to be true.

Gödel's Incompleteness Theorem has been used to argue that we may never be able to fully understand ourselves.

That was exactly what Russell and Whitehead had in mind with *Principia Mathematica*, although they weren't alone. Giuseppe Peano established five axioms for doing arithmetic on the natural numbers in the 1890s, and Zermelo and Fraenkel laid down the basics of set theory in 1908 — both of these systems are still widely used in serious math today.

However, they're broken. All of them. As is every possible mathematical system you can come up with. Kurt Gödel proved this in 1931, reasoning as follows: every true statement in a formal system can be encoded as

Abram Halevi Fraenkel (1891–1965), German-born Israeli mathematician.

Ernst Friedrich Ferdinand Zermelo (1871–1953), German logician and mathematician.

Kurt Gödel (1906–1978) was an Austrian-American logician, mathematician and philosopher.

a whole number — an enormous, complicated number.

True statements can be generated from other true statements by applying certain rules to these numbers — again, enormous, complicated rules.

Gödel asked what would happen if you encoded a statement equivalent to "This statement cannot be proved within this system."

It's Epimenides's paradox all over again: if the statement is true, then it cannot be proved, and so the system isn't complete.

If the statement is false, then it can be proved, so it's both true and false — so the system isn't consistent. Your only alternative is to say the statement is undecidable — so the system isn't decidable. There is one subtle difference to the typical Epimenides-style paradox: "cannot be proved" is used in place of "false" — Tarski showed that this replacement was necessary.

If you adopt this statement as an axiom, you can apply the same reasoning to the resulting system, and end up with another system-breaking sentence.

It's an absolute doozy of a proof. Any formal system is broken in one of those three specific ways: completeness, consistency and decidability.

Math took the sensible option and decided that decidability had to go, accepting that there were some statements that could be neither proved nor disproved.

Examples include the "continuum hypothesis", which states "there is an infinity with cardinality between that

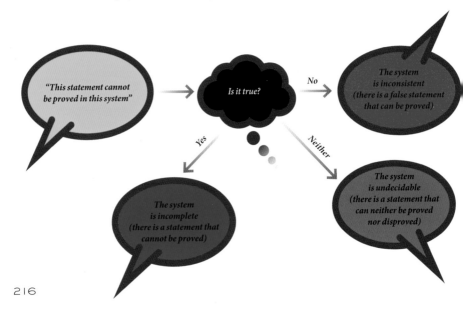

Paul Cohen was awarded the Fields Medal (the "Nobel Prize" of math) in 1966 for his work on the axiom of choice.

of the natural numbers and that of the real numbers"; and the "axiom of choice," which states that, for any set of nonempty sets that have no members in common with each other, there exists at least one set that contains exactly one member from each of these sets. The *axiom of choice* was finally proved undecidable by Paul Cohen in the 1960s, following Gödel's work.

> *The supreme triumph of reason is to cast doubt upon its own validity.*
> Miguel de Unamuno

KURT GÖDEL

If Évariste Galois had the greatest romantic death in mathematical history, the Austrian Kurt Gödel's is probably the most tragic.

Born in Brünn, Austria-Hungary (now Brno in the Czech Republic), Gödel studied mathematics in Vienna, although by the time he enrolled, he was already proficient with university-level math. He took part in a seminar run by Moritz Schlick based on Russell's *Introduction to Mathematical Philosophy*, and became hooked on mathematical logic. A lecture by Hilbert inspired him to begin study of the completeness of formal systems. His doctoral thesis proved that a relatively simple form of mathematical logic, the first-order predicate calculus, was complete: everything that was a) expressible in the system, and b) true, could be proved using its rules.

He went on to take the *Principia Mathematica* to pieces, explaining not only that it was wrong and why, but proving that no such system could ever be complete, consistent and decidable.

Before the outbreak of World War II, Schlick was assassinated and Gödel had a breakdown. He developed paranoid tendencies and was hospitalized. Despite this, he was found fit to serve in the German military. He escaped via the Trans-Siberian railway, and joined the Institute of Advanced Study in Princeton,

Friedrich Albert Moritz Schlick (1882–1936) ran the seminar that got Gödel hooked on mathematical logic. He was murdered by a former student.

which he had visited several times in the 1930s and made friends with Einstein.

He became an American citizen in 1947, despite almost blowing the interview. He thought he'd discovered a flaw in the constitution and — rather than give the expected answer "no, sir" when asked whether a Nazi-style dictatorship could ever come to pass in America, he started to elaborate. Luckily, the sympathetic judge steered the subject to something more routine.

Gödel shared the first Albert Einstein Award with Julian Schwinger in 1951 and was awarded the National Medal of Science in 1974.

It was his paranoia that did him in, in the end. He had a morbid fear of being poisoned and the only person he trusted to cook for him was his wife, Adele. In 1977, Adele was taken to hospital. Early in 1978, Gödel starved to death. Poor guy.

Princeton Institute of Advanced Study became Gödel's haven when he fled the Nazis.

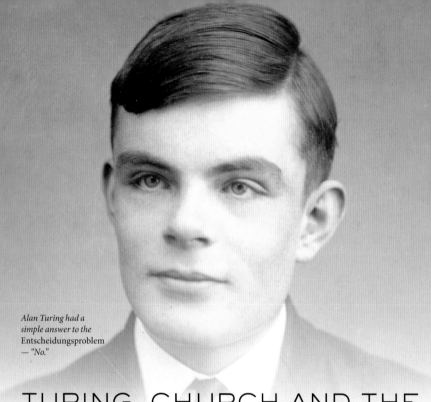

Alan Turing had a simple answer to the Entscheidungsproblem — "No."

TURING, CHURCH AND THE ENTSCHEIDUNGSPROBLEM

Twenty-three problems weren't enough for Hilbert. He kept adding to his "program," eventually borrowing a question roughly posed by Leibniz two centuries earlier.

It is generally known by its German name, the *Entscheidungsproblem*, which translates as "decision problem" — a mouthful of a name if ever I heard one. It asks if there is an algorithm that takes a statement of first-order logic and tells you whether or not it's true. The biggest problem was deciding what constituted an algorithm.

In 1936, both Alonzo Church (with something called the lambda "λ" calculus) and Alan Turing (with an idealized computer called a Turing machine) independently came up with formalizations of algorithms — when Church's version was brought to his attention, Turing noted that the two were equivalent;

both concluded that the answer to the *Entscheidungsproblem* was "no."

They did this by drawing on Gödel's work. Each system of algorithms was exactly the kind of formal system Gödel's Incompleteness Theorem could be applied to — meaning that they can be broken in exactly the same way as any other formal system.

So, not only is math broken, so is your computer. Sorry. Buying a new one won't help, either.

Alonzo Church, invented λ-calculus.

$$\lambda$$

$$f^n = f \circ f \ldots \circ f.$$
$$\underbrace{}_{}$$
$$n \text{ times}$$

Number	Function definition	Lambda expression
0	$0\ f\ x = x$	$0 = \lambda f.\lambda x.x$
1	$1\ f\ x = f\ x$	$1 = \lambda f.\lambda x.f\ x$
2	$2\ f\ x = f\ (f\ x)$	$2 = \lambda f.\lambda x.f\ (f\ x)$
3	$3\ f\ x = f\ (f\ (f\ x))$	$3 = \lambda f.\lambda x.f\ (f\ (f\ x))$
...		
n	$n\ f\ x = f^n\ x$	$n = \lambda f.\lambda x.f^n\ x$

Lambda calculus provides a form of computation that, in its idealized form, can perform any calculation. It forms a key part of programing theory, and is thought of as the smallest universal programing language. In the above table, the function f is applied to the argument x the number of times specified by the number on the left.

BABBAGE, LOVELACE AND THE DIFFERENCE ENGINE

The precise ancestry of the computer is in some dispute — depending on who you listen to, the parents of computing might be Pascal, Leibniz, Jacquard, Turing, Hopper, or any one of several others.

One person with a good claim to a branch of the family tree is Charles Babbage (1791–1871).

Like most of the mathematicians featured here, Babbage was a little on the eccentric side — he gained notoriety for his campaigns against street music (as a result, organ-grinders would frequently set up stall outside his windows, just out of spite) and against the children's game of hoop-rolling, on the grounds that the hoops kept tripping up the horses.

Also like many of this book's inventive heroes, he wasn't limited to the thing he was famous for: he invented the ophthalmoscope for looking inside the eye (although his

Charles Babbage, father of the computer.

invention was ignored) and the cow-catcher, used by trains to clear their paths of obstacles; he also cracked the Vigenère cipher before Kasiski did (although, owing to military secrecy, this wasn't established until the 1980s). He was also responsible for the eventual adoption of Leibniz's notation for calculus in Britain, establishing the British Lagrangian School in an effort to help British science catch back up with the Europeans.

His biggest accomplishment had its seed in 1812, when he heard that the French government had instigated a project we'd probably call parallel computing today: in order to

Babbage hated the street music of the barrel organ, but the "punch card"-type system it used was an interesting idea …

calculate the correct values in a table of logarithms, they split the task up into small, simple calculations that could be done by hand, then employed a large team of human computers to do the work. Babbage, a keen industrialist, asked himself: why bother with the humans?

A working model of Babbage's computer was finally built in the early 1990s.

It wasn't until a decade later that Babbage started on the project of the Difference Engine, a two-meter tall, fifteen-tonne, steam-powered behemoth that would calculate the value of certain functions without the need for human intervention. Sadly, it was never completed — for reasons unclear to me,

Babbage didn't spend the adequate grant money he received from the Government on finishing the machine, and eventually, the cash dried up. All the same, the plans were solid — the Science Museum in London built a working model of this mechanical computer in the early 1990s.

However, the Difference Engine isn't a computer any more than the Pascaline or Leibniz's mechanical calculator (don't get me wrong — it's a fantastic contraption — it's just not a computer). Luckily, he invented one of those, too.

Babbage's follow-up project, the Analytical Engine didn't just do sums — it followed instructions on punch cards, inspired by the Jacquard loom. It had control flow, so you could execute loops, conditional statements and subroutines. It had memory, so you could store values. It was mathematically equivalent to a modern computer, although perhaps a teeny bit less powerful. Or at least, it would have been, had it ever been built. His failure to produce a working

Difference Engine counted against him, and he never received funding to build it.

That didn't prevent the Analytical Engine being studied, though. In 1842, Luigi Menabrea wrote a description of the engine in French, which Babbage's assistant, Ada Lovelace (1815–1852) translated and annotated extensively — including an algorithm for the computation of Bernoulli numbers, which is widely considered the first computer program. The computer language Ada is named in her honor.

Ada Lovelace was Charles Babbage's assistant and can be regarded as the first computer programer.

Grace Hopper was a pioneer of computer technology and one of the first programmers of electronic computers.

GRACE HOPPER

If you go to YouTube, you can see Rear Admiral Grace Hopper (1906–1992) giving a presentation, holding up about 12 inches of electrical cable and describing to the assembled room that what she has in her hand is a nanosecond.

That's how long it takes an electrical signal to travel through that much cable at the speed of light. She's an elderly lady — judging by the haircuts in the room, it's from the 1980s — and she's firmly, humorously explaining how to use these to show "husbands, children… Admirals, Generals, people like that" how long it takes signals to move around.

She seems physically frail, but it's perfectly clear that this is not someone that you would want to mess with. For someone showing such brilliant disrespect to authority, she's commanding an awful lot of it.

And so she should: she was one of the first programers of an electronic computer (the Harvard Mark I, in 1944), the first to write a compiler (a program that takes high-level instructions like "work out the sine of 20°" and turns it into something a computer can execute directly) and was closely involved in the development of COBOL in the late 1950s, one of the first programing languages that could be read as natural language.

She's also remembered for introducing an engineering term

The first "Computer Bug," after it was removed from the Harvard Mark II computer in 1947.

into computer science. It's documented in a logbook: on September 9, 1947, the Harvard Mark II computer had stopped working properly. One of the engineers determined the cause of the problem: a moth had got caught in one of the relays. Hopper dryly noted that she had "debugged" the computer — and the process of getting rid of glitches in programs is still called debugging today .

My favorite description of Hopper comes from Jay Elliot, once Senior Vice President of Apple Computer:

"[she] appears to be 'all Navy,' but when you reach inside, you find a 'Pirate' dying to be released."

Young people come to me and say, "Do you think we can do this?" I say, "Try it!" … I stir 'em up at intervals so they don't forget to take chances.

Grace Hopper

CHAPTER 10
HOW WE WRITE IT

The language of mathematics develops from tally marks to computer code.

Computer codes may look simple, but it's taken us a long time to arrive at the symbols to write it down easily.

ROMAN NUMERALS

A Roman centurion walks into a bar, holds up two fingers and says,
"Five beers, please."

Roman numerals are, frankly, a terrible way to write down numbers. There's a certain logic to them, although it's the same logic that decides what denominations of coins and notes your country's mint produces: it's arbitrary and sometimes frustrating.

They look like letters, although you can make a good case for the system developing out of tally marks and the mnemonic patterns in them, so perhaps they're symbols.

Whatever they are, they're assigned to certain numbers: I is 1, V is 5, X is 10, L is 50, C is 100, D is 500 and M is 1,000. In its simplest form, you can build other numbers by combining the symbols, with the largest number first:

1,751 is MDCCLI –
1,000 + 500 + 100 + 100 + 50 + 1

The Romans conquered much of the known world, organized huge armies and built impressive cities, despite possessing an appallingly unwieldy number system.

The Romans were nothing if not lazy, though, and objected to having to carve more than three identical symbols in a row. Instead, they would write a smaller symbol in front of a larger symbol to show that that value was taken off: IV would be one taken away from five (four). XC would be 10 taken away from 100 (90).

Roman numerals still crop up on many clock faces.

This does make significant savings for something like 1,999 — without abbreviations, it would be MDCCCCLXXXXVIIII, but the more sensible way of writing it would be MIM. Adding up and taking away with Roman numerals is fairly straightforward. It's not *all* that different to how you add up Arabic numerals, but you hit a problem if you want to multiply.

I don't know where you'd start if you wanted to multiply VII by IX, at least not without cheating and thinking about more modern notation. If you were a Roman with a multiplication problem you needed to do, you might use a Babylonian-inspired abacus or a reckoning board using loose pebbles.

The Latin word for pebble, incidentally, is *calx* — "calculation" was originally pushing around stones to do a sum. The word also leads to calcium and (more importantly), calculus.

Dividing is also difficult (except, possibly, if you're dividing by precisely a symbol in the system — dividing by X or V isn't too hard). The Roman system did have fractions — twelfths were represented by dots (the Latin name for a twelfth, *uncia*, is where the words "ounce" and "inch" come from), and halves by the letter S.

This system stuck around in Europe for a millennium or so — and change, when it came, was controversial.

THE ABACUS

The first calculating machine (unless you count tally sticks, which I don't), wasn't quite an abacus, but it had the same sort of idea. The Sumerians, around 2500 BCE, used tables with columns representing the various strands of their base-60 number system to add and subtract. Slightly later, the Egyptians used pebbles to represent the numbers being used, a system later used by the Persians and the Greeks.

The Chinese suànpán *was the earliest abacus.*

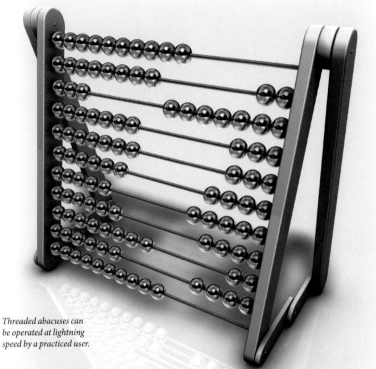

Threaded abacuses can be operated at lightning speed by a practiced user.

The Chinese *suànpán*, though, was probably the first abacus we'd recognize today: rather than running in grooves, the counters are threaded on rods — five in the lower half (which each count for one unit), and two up top (which each count for five). Really, you only need four below and one up top for a base-10 system — although a five-and-two layout makes it possible to work in hexadecimal, or even octadecimal if you're so inclined.

Unlike the grooved versions, the threaded versions are quick and easy to manipulate, and offer quick algorithms for the four main operations, as well as square and cube roots.

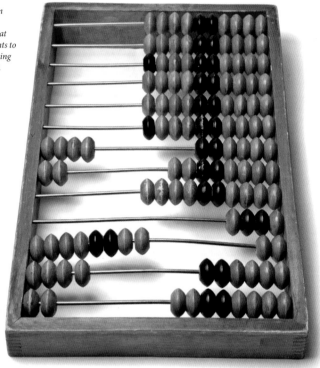

The Russian abacus has dark beads at certain points to make counting them easier.

The Russian abacus is much more familiar, having 10 beads on a wire. To make it quicker to read, the two beads in the middle of each wire are darker than the others, making it simple to tell the difference between (say) 7 and 8 without having to count.

Similarly, the final beads on the "thousand" and "million" wires are a different color, so you can read off your numbers without losing count. This is much the same as using commas to separate off thousands and millions in a decimal number.

Today, in most schools, the abacus is a curiosity, although they're an excellent tool for explaining methods of arithmetic. Schools across Asia use them as a tool for developing mental arithmetic skills — after a while, some students become so proficient they no longer need a physical abacus and can simply visualize the beads as they move, often at lightning speed. A champion user of the mental abacus can add 15 three-digit numbers in well under 2 seconds — much quicker than is possible on a calculator.

However, there's one type of school in which the abacus is invaluable. If you're blind, the stages of writing and reading back workings for arithmetical calculations are extremely difficult, even using Braille and Nemeth code.

You have two choices if you want to learn math: you may use a talking calculator, or you may use an abacus. Only one of those teaches you the methods you need! A *Cranmer abacus*, specially designed for use by the blind, has beads that are harder to move accidentally and won't slide out of place.

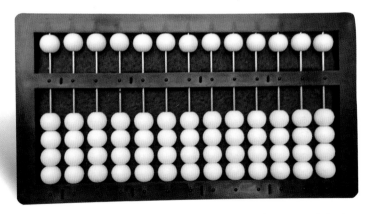

This Cranmer abacus has been modified for use by a blind person. A felt lining stops the beads from slipping, while raised dots on the cross bar allow the user to feel which column is which.

MUCH ADO
ABOUT NOTHING

For something that isn't even there, the number zero has caused an awful lot of confusion over its fairly short lifespan.

In terms of human thought, zero is much younger than the other numbers — number systems did very happily without it, thank you very much, until at least 400 BCE, many centuries after counting began.

Part of the trouble is that zero really represents two different things: the absence of stuff (as in, I own zero cows)

We are so familiar with zero, used as a basis of computer binary code with its zeros and ones, that it's hard to imagine a world in which it does not exist.

or as a placeholder in a written-out number, so you can tell the difference between 85, 850 and 805.

The second of these was in place in the late Babylonian era — 805 would be represented effectively by 8"5, with a double-wedge symbol showing that there was a gap between the hundreds and the units. Actually, the Babylonians

The first recorded use of 0 is at Gwalior in Madhya Pradesh, India.

degree by Ptolemy around 130 CE (the Greeks may well have introduced the symbol 0), but it was in India that the idea really took hold. Mathematicians understood the importance of showing "gaps" between numbers, but didn't have the notation for it. At first, they used dots to represent the spaces.

used a base-60 system, so it was really between the 3,600s and the units, but the idea still holds. There was still, however, no difference in notation between 85 and 850 — you'd have to rely on the context to work out which was meant.

Zero as a digit was developed to some

The first dated use of 0 as a digit in its own right is from 876 CE — a garden in Gwalior, in Madhya Pradesh, India is described as 187 by 270 hastas, and producing 50 garlands of flowers a day — all of which numbers are written almost exactly as you'd write them today.

The Gwalior zero is rather later than the introduction of zero as a usable concept. In the 7th century, the Indian mathematician and astrologer Brahmagupta worked out the basic rules of arithmetic using zero — if you add it or subtract it, it doesn't change anything;

The Indian mathematician Bhaskara II was born in the Bijapur district of Karnataka, where the Gol Gumbaz temple stands today.

if you multiply by it or divide it by something, you're left with nothing.

However, he believed that dividing by zero was possible, and just gave you a fraction like $7/0$. Five hundred years later, another Indian mathematician and astronomer, Bhaskara II suggested that $n/0$ was infinity — although this would lead to all manner of paradoxes.

It's difficult for a mathematician today to understand how math could be practiced for millennia without the use of a zero (zero was still unconventional by the time of Cardan and only really took hold after the Renaissance) and yet, it was. It's not exactly fair to point out that the bulk of the major advances in math took place after that, but I'm going to do it anyway.

NONE MORE EQUAL

Mathematicians, in general, are lazy. Give us an hour to solve a problem, we'll spend the first 55 minutes looking for a way to do it in 5 minutes. And when we write something down, we write it down tersely. There's no room for rhetoric.

For example, take the quadratic formula:

If:

$$ax^2 + bx + c = 0$$

Then:

$$x = \frac{-b \pm \sqrt{b^2 - 4ac}}{2a}$$

That says, if you know that the square of a number, multiplied by some number, added to the original number multiplied by some other (although it could be the same) number, added to a third number, gives you a total of zero, then you can work out the original number.

All you need to do is take the second number and square it, subtract four times the product of the first and third numbers, then square root the result.

Add that to or subtract it from the negative of the second number and divide the whole lot by double the first number and boom! There are your answers.

Nicole Oresme was the Dean at Rouen Cathedral in France.

However, for a substantial part of the history of math, that's pretty much how equations were written down. The symbol for addition, +, didn't come into use until the 14th century, possibly introduced by Nicole Oresme as a contraction of the word "et," meaning "and." It was more common to use the abbreviations p and m to mean plus and minus. Some authors wrote a tilde (˜) over the m to show it meant minus rather than just the letter m and, eventually, the m itself was dropped in favor of the tilde on its own. It straightened out over time to become a line like this: —

The greatest bit of laziness, though, is due to the Welsh polymath, Robert Recorde (1512–1558) in his brilliantly titled *Whetstone of Witte*. He says, "to avoid the tedious repetition of these words 'is equal to,' I will set ... a pair of parallels, or Gemowe [twin] lines of one length (thus =), because no two things can be more equal."

Incidentally, the title is a play on words: algebra was known as the "cosslike practice" — and *cos* is Latin for whetstone, neatly suggesting that algebra sharpens the mind. The equals sign wasn't an immediate hit — some preferred || to mean equal, whereas others stuck to the abbreviation æ for *æqualis*, the Latin for equals.

Robert Recorde was a fellow of All Souls College, Oxford.

> *All men are equal, but some are more equal than others.*
>
> George Orwell

English mathematician John Wallis (1616–1703) introduced the symbol for infinity.

Up to around 1600, progress in mathematical symbols had been painfully slow: half a dozen new symbols over the course of 250 years (the plus and the minus, the radical and the parentheses, and finally the equals sign).

At that point, the invention of symbols went a bit crazy. Before getting too far into the 18th century, we had:

- The multiplication sign and a symbol :: for proportion, dreamed up by William Oughtred — as well as the abbreviations sin and cos.
- The infinity symbol, introduced by John Wallis.
- ÷ for division, courtesy of Johann Rahn.
- π (the first letter of the Greek word for perimeter) for the ratio between the diameter and circumference of a circle — suggested by William Jones, but not widely adopted until Euler started using it.

On the whole, mathematicians were fairly careful not to introduce symbols that interfered with letters they may use as variables. Leibniz slipped up a little bit by using "d" for differentiation although, at least in typeset math, the d is usually in roman font rather than italics (*d*) like a variable.

Euler also erred with "e" and, for that matter, "f" but given that functions can, in some contexts, be used as variables, it probably didn't matter too much.

Mathematical notation developed haphazardly over the course of several centuries, at the whim and convenience of the people writing, which means that the conventions of today are an unholy mess. There's not even worldwide agreement on what symbol to use as a decimal point: in the English-speaking world, it's a dot; elsewhere, it's a comma.

Unfortunately, like English spelling and the QWERTY keyboard, it's too firmly entrenched to change, even if better alternatives exist.

Or is it?

Slow and unwieldy, the QWERTY keyboard endures because we are familiar with it and don't like change.

POLISH (AND REVERSE POLISH) NOTATION

Of course, proper geeks — really, truly, proper geeks — disdain calculators completely. Who needs them, when mental arithmetic is so easy?

However, for those who do condescend to stoop to the use of machines to help them, there's only one sort of calculator worth buying: one that uses Reverse Polish Notation, or RPN.

Instead of mangling buttons to write …

$$\left(\frac{4\pi+7}{3}\right)^{\pi^2}$$

… an RPN user might type:

$$4\,\pi\times 7 + 3 \div \pi\,2\;\wedge\;\wedge$$

Less legible? Perhaps. But look, ma — no brackets!

That was what inspired Jan Łukasiewicz (1878–1956) to introduce what became known as Polish Notation in 1924.

As long as you know how many operands each function takes (its *arity*: +, for example, has an arity of two because you add two numbers together; a function like sine has an arity of one, as it only takes one argument), a Polish Notation "sentence" is perfectly well defined.

The only difference between Polish and Reverse Polish is that in RPN, the function comes after the arguments rather than before — which is a bit more efficient.

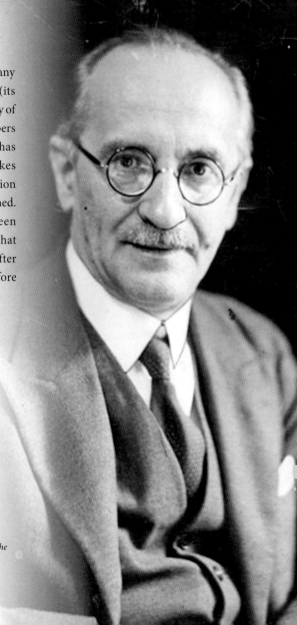

Polish logician and philosopher Jan Łukasiewicz was a professor at Lvov and Warsaw Universities. After World War II, he lived in exile in Dublin.

SO HOW DOES REVERSE POLISH NOTATION WORK?

In order to use Reverse Polish Notation, you work through the sentence from left to right, just like reading English.

If you get a number, you add it to the *stack*; if you have a function, you apply it to the appropriate number of things at the end of the stack.

$$4\,\pi \times 7 + 3 \div \pi\, 2\, {}^\wedge{}^\wedge$$

In this example, you would start by pushing 4 and π onto the stack. [2]

The × means multiply the last two things together, and replace them with the result: the stack now just has the element 4π (which is about 12.6). [3]

You then add 7 to the stack. [4]

The + means add the last two things on the stack together and replace them with the result — so you have $4\pi + 7$, which is about 19.6. [5]

Then add 3 to the stack [6], then divide the last two numbers (about 6.5). [7]

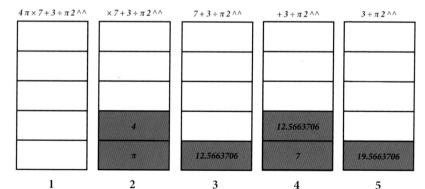

Next, add the π and the 2 to the stack, which is now 6.5, π, 2. [8]

The caret means raise the penultimate number to the power of the last number in the stack and replace them — is about 9.9, making your stack 6.5, 9.9. [9]

Lastly, raise the penultimate number to the power of the last number, giving 110,000,000 or so. [10]

That's exactly the same thing you do — and the same answer you would get — from the mess of fractions and brackets on the previous spread.

Among the many advantages of RPN — not counting the fact that it's much easier for computers to understand and parse than traditional notation — are that you don't need to remember the order of operations: you just do them in the order they come along.

If I was redesigning how math is written (and frankly, it could do with a revamp), I'd certainly look at RPN as a good place to start.

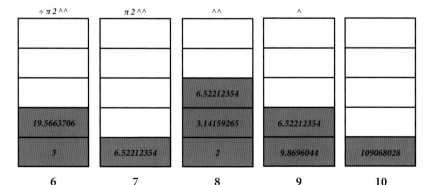

CHAPTER 11
THE SCOTTISH CAFÉ

*A professor interrupts a late-night argument in a Kraków park, a notebook
is purchased to save the furniture (and preserve what would have been written on it),
a live goose is presented on Polish TV and a camel becomes rather confused.*

*Professor Hugo Steinhaus was astounded by what
he overheard during an evening stroll in the park.*

PLACE: LVIV (OR LWÓW)

The city of Lviv lies in the west of modern-day Ukraine, close to the borders with Slovakia and Poland.

Since the start of the 20th century, it's been, at various times, the Polish city of Lwów (because it was called Lwów at the time this chapter takes place, I call it Lwów throughout) and the Austro-Hungarian city of Lemberg (not to mention the Soviet city of Lviv) — none of which are as pretty a name as the more romantic Latin version, Leopolis: the city of lions.

Between the two World Wars, Lwów was Poland's third-largest city, and one of its most vibrant cultural and academic centers. Sitting at a crossroads between the Soviet Union and Western Europe, it was an important trade hub.

Lwów University is among the oldest in Europe.

Its university is among the oldest in Europe (founded 1608), and in 1901, the Lwów Scientific Society made its home there, although it was originally called the Association of Support of Polish Sciences, changing its name in 1920.

The society's members included Rudolf Weigl, who developed the first effective antityphus vaccine; Henryk Actowski, one of the first scientists to spend a winter in Antarctica; and the mathematician Stefan Banach.

It's Banach we're interested in — and his habit of going to the local café to talk math with his colleagues.

It's not clear whether the landlord was upset about the proofs they kept writing on his tables, or if the mathematicians were upset about them being wiped off at the end of the day.

It's not even clear whose idea the notebook was, but I like to think it was Łucya Banach, whose husband was one of the key figures in the Scottish Café.

Memorial to Polish mathematician Stefan Banach in Krakow, Poland.

The site of the Scottish Café in Lwów as it looks today.

The building is still there, but the chess boards and the coffee pots are long gone — it's now a branch of the Universal Bank. But for nearly a decade, if you wanted to find a *great* mathematician in Lwów, you would look at the Scottish Café. Time your visit between 5 and 7 pm, and look for the people yelling at each other at the tables in the middle of the room.

Almost inevitably, you would find the great mathematician Hugo Steinhaus, arguably the founder of what we know as functional analysis.

That's not where they started gathering. Their first meeting place was at the Roman Café. They'd play chess, drink coffee and beer, and argue about the mathematical problems of the day.

The mathematicians, after one disagreement too many, possibly about the furniture, possibly about credit, soon moved over the road to the Scottish Café.

You would also find his student, Stefan Banach, so brilliant that the academic authorities simply could not ignore him. Despite his refusal to take any exams, he still somehow ended up with a doctorate.

Then there was Stanisław Ulam, who later joined the Manhattan Project; Mark Kac, famous for his work in spectral theory deciding whether you can infer

Mathematicians gathered in Lwów in 1930.

1) L. Chwistek, 2) S. Banach, 3) S. Loria,
4) K. Kuratowski, 5) S. Kaczmarz, 6) J. P. Schauder,
7) M. Stark, 8) K. Borsuk, 9) E. Marczewski,
10) S. Ulam, 11) A. Zawadzki, 12) E. Otto,
13) W. Zonn, 14) M. Puchalik, 15) K. Szpunar

the shape of a drum from its sound (generally, no, you can't); Karol Borsuk, a creative topologist; Stefan Kaczmarz, whose work led to CAT scans; Bronislaw Knaster, best-known for devising (with Steinhaus and Banach) a fair way to divide cake; Stanislaw Saks, who wrote *The Theory Of The Integral* before joining the Polish underground during World War II; and Stanislaw Mazur, who ended up presenting fellow mathematician Per Enflo with a live goose on Polish TV.

THE SCOTTISH BOOK

Like a handful of the problems in the notebook Mrs. Banach bought, Problem 153 has a bounty attached: Mazur promised a live goose to anyone who could solve it.

In Depression-era Poland, the prizes weren't always as extravagant as a goose — two small beers, a bottle of wine, a demitasse of coffee — but for solving one of the harder, later problems, you might win 100 g of caviar, lunch at the Dorothy in Cambridge, a fondue in Geneva or — courtesy of the legendary John von Neumann — "a bottle of whisky of measure > 0."

The book itself was kept at the café. Anyone could ask to see it, but (in principle, at least), questions would only be written on the left-hand pages after the group had devoted considerable time to them, although a few less difficult questions crept in. Solutions, if solutions could be found, ended up on the facing page.

In among the maze of *metric spaces*, Knaster-Kuratowski-Mazurkiewicz maps (which, I think, are just as terrifying as they sound), Lebesgue measures and Lipschitz conditions, there are a few problems that can be set up in more-or-less clear language.

Problem 38, for instance, asks what happens if you have N people, each with k followers chosen at random — what's the probability of being able to connect every person to everyone else by a chain of mutual followers if N is very large?

It turns out that if k is at least two, you can almost certainly connect everyone, but if k is set at only one, you almost certainly can't.

Part of The Scottish Book.

Linking people together in unusual ways was the crux of The Scottish Book's *Problem 38.*

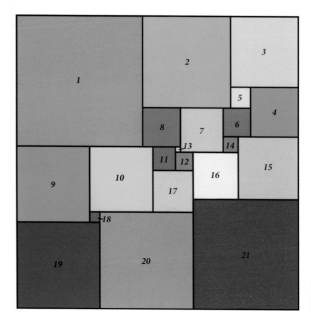

One possible solution to Problem 59 using only 21 squares.

Problem 59, by Ruziewicz, asks whether you can split a square into a finite number of smaller squares, all of them different sizes.

This was solved in 1940 by Ronald Sprague, who found a solution with 55 small squares.

As is typical with math research, this asked as many questions as it answered: was this the only solution? What's the smallest number of squares you need? It's not the only solution — R.L. Brooks found one with just 26 squares later that year, but even that wasn't as low as you can possibly go.

A.J.W. Duivesteijn proved that you'd need at least 21 squares in the 1960s, but didn't find a 21-square example until 1978 — nearly 40 years after the question was posed.

STEVEN BANACH

One evening in 1917, Professor Hugo Steinhaus took a stroll through a Kraków park, and overheard a heated conversation on a park bench.

Not your usual evening argument in the park, this: the two men were discussing the details of the Lebesgue integral, something even Steinhaus had only recently come across. One of the men was Stefan Banach (1892–1945), who astonished Steinhaus with his ability to tackle extremely difficult problems.

It wasn't the only astonishing thing about Banach. He refused to take exams, but received a doctorate anyway — his numerous papers and ideas convinced the university to waive the usual rules. In the 1920s, Banach gained a professorship at Lwów University and went on to found the Lwów School of Mathematics.

Excluded from holding a university post after Lwów was overrun by German forces in 1941, Banach was reduced to feeding lice in Rudolf Weigl's laboratory until Russian forces took back the city in 1944, at which point Banach planned to return to Kraków. He died the following year after a short battle with lung cancer.

His monograph *Theory of Linear Operations* is considered the foundational text of functional analysis. As one of the most influential mathematicians of the 20th century, he's known for many things, but particularly Banach spaces (an especially well-behaved kind of vector space) and the Banach-Tarski paradox.

Warsaw was left in ruins at the end of World War II.

PER ENFLO
AND THE GOOSE

Problem 153.

Given a continuous function f(x,y) defined for $0 \leq x, y \leq 1$ and the number e > 0; do there exist numbers $a_1 \dots a_n, b_1 \dots b_n, c_1 \dots c_n$ with the property that

$$|f(x, y) - \sum_{k=1}^{n} c_k f(a_k, y) f(x, b_k)| < e$$

Prize: A live goose.

Mazur, November 6, 1936

Unless you're pretty hot on your functional analysis — and translating Problem 153 from Polish to English doesn't help a great deal — these days, it's usually phrased as: "Does a separable Banach space have a Schauder basis?"

This is typically the way with analysis: — you can express a problem in nine words, two of which are proper nouns and seven of which are fairly common English words, yet unless you know what a Banach space and a Schauder basis are,

you're stuck. So you look up a Banach space — it's a complete, normed vector space. Now you have three more things to look up.

Unlike most of the problems in the Scottish Book, problem 153 didn't yield a quick solution.

Legend has it that, for the bulk of the problems, once they were written down, Banach would go home and come back to the café the next day with an outline of the proof. This one, however, was different. Apart from realizing it was closely related to the Schauder basis, there wasn't much forward movement for over 35 years.

Except for one thing: in 1955, Alexander Grothendieck showed that problem 153 was equivalent to the Approximation Problem. This effectively served to rewrite the problem as "Does every Banach space have the approximation property?"

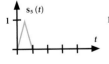
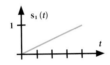
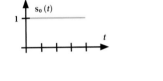
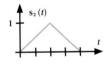

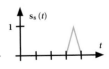
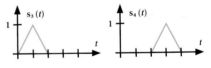

An example of a Shauder basis: Any continuous function on [0,1] can be made by adding together multiples of these functions. They go on forever and the next line would comprise eight narrower peaks, then 16 and so on.

A live goose was the star of the show when two mathematicians met on TV.

As well as winning a goose, Per Enflo developed new techniques for use in functional analysis.

The answer, it turns out, is no. In 1972, Swedish mathematician Per Enflo contrived to construct a Banach space with neither a Schauder basis nor the approximation property, solving both problems and problem 153 at once.

Mazur was not the kind of person to let the passage of 36 years get in the way of a bargain. He had offered a bounty to the solver of the problem, and Enflo was entitled to his prize.

TV audiences in Poland were treated to the spectacle of one mathematician handing a live goose to another in an elaborate ceremony.

Stan Ulam escaped to America with his family just 2 weeks before the Germans invaded Poland.

STAN ULAM

Stanisław Ulam (1909–1984) was born in Lwów, and a member of the Lwów School of Mathematics — of the nearly 200 problems in the Scottish Book, Ulam was credited on over a third. He was also responsible, after World War II, for translating a surviving copy into English.

Ulam split his time in the 1930s between Poland and working with John von Neuman at the Institute of Advanced Studies in Princeton. His family left for America in 1939, 2 weeks before the German invasion.

In 1943, Ulam was invited to join the top secret Manhattan Project in New Mexico — he deduced what the project was after checking a guidebook to the area out of the library, and noticed that the last few people to read it were all physics professors who had then mysteriously disappeared.

While recovering from an attack of encephalitis in 1946, Ulam played a lot of solitaire. Being a mathematician, he wondered what the probability of winning the game was, and realized he could estimate it by playing a huge number of games and seeing how often he won — and he could use statistics to work out how accurate his estimate was. This was the birth of the Monte Carlo simulation: instead of trying to find elegant solutions to tricky problems, you let a computer simulate things under random conditions, and see what the average result is.

Even with the limited computing power available at Los Alamos, it was a more efficient method for solving the problems his team was working on than traditional methods. Von Neumann and Metropolis first used the method to study neutron diffusion in 1947.

Cpl. Irwin Goldstein (foreground) sets the switches on one of the ENIAC's function tables. The world's first electronic computer was used to run calculations for the hydrogen bomb during World War II.

Ulam is also responsible for the Ulam Spiral. Killing time in a dull lecture in 1963, he started doodling numbers in a spiral, and coloring in the primes. To his surprise, they tended to fall on diagonals, meaning that — although there's no formula for generating primes — some simple equations, such as $P = x^2 - x + 41$, generate primes much more frequently than expected.

Ulam wasn't the first to explore such patterns — some 30 years earlier, Laurence M. Klauber had noticed something similar involving a triangular grid, and Arthur C. Clarke had mentioned the Prime Spiral in 1956. He didn't actually draw it out, or else we'd probably be talking about the Clarke Spiral instead of the Ulam spiral!

Ulam died from a heart attack in 1984.

Do not lose your faith. A mighty fortress is our mathematics. Mathematics will rise to the challenge, as it always has.

Stanislaw Ulam

Ulam spiral using all numbers.

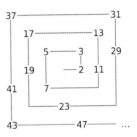

Ulam spiral using prime numbers.

Ulam spiral showing patterns of prime and composite numbers.

THE END OF THE SCOTTISH BOOK

The turmoil of World War II brought an end to contributions to The Scottish Book. The book was spirited away after the Nazi invasion, and was found in Banach's effects after his death in 1945.

Banach's son showed it to Steinhaus, who arranged for a typewritten copy to reach Ulam — by now at Los Alamos — in the mid-1950s. Ulam had it typed up and distributed 300 copies at his own expense. Requests for it continued to come in, and a conference to discuss it was arranged. This took place in 1979 at North Texas State University, by which time around three-quarters of the problems had been solved. New Scottish Books had started springing up the world over. I hope many more do.

A plan was hatched to bury The Scottish Book *under a set of football goal posts for safekeeping when the Nazis invaded. It survived unburied.*

Taverns and coffee houses have a long history as venues for debate.

MATHJAM

Walk into The Jam Factory in Oxford, or the Grape and Olive in Cardiff, or dozens of other pubs around the world, on the penultimate Tuesday evening of the month, and you may see scenes reminiscent of the Scottish Café in its heyday: a group of excitable mathematicians, ranging from students to retired professors, sitting around tables playing games, working on puzzles, and arguing about math problems. It's a MathJam.

Although MathJams tend to cluster around university towns, it's not an academic organization, and anyone who's at all interested in math is welcome. The promotional material says the gatherings are meant to be accessible to anyone "down to, and including, chemists."

In reaction to the fact that most mathematical clubs are explicitly educational — get your basic math up to speed, or study for your exams — MathJams are held in pubs, in the evening, and aimed at doing math for its own sake.

Nobody has yet offered a live goose as the solution to a problem, but that's surely only a matter of time.

For more information on your nearest MathJam, or on setting one up, visit mathjam.com.

THE BANACH-
TARSKI PARADOX

*Take a ball — a mathematical ball, one you can hold in your head. Cut it up into pieces
— five is plenty. Reassemble them, and voilà! Two balls, each identical to the first!*

This is the *Banach-Tarski Paradox*: you can create something from nothing! It sounds like a cheap trick, at first — maybe the balls are hollow, or they're different sizes or something — but no, it's mathematically legitimate, as long as you accept the Axiom of Choice.

In principle, if a golf ball were a mathematical object, you could split it into a finite number of pieces and reassemble them to make the Great Pyramid of Cheops.

So, you can take a ball of gold and a sharp knife and end up with more gold than you started with? Sadly, not. For a start, mathematical objects can be cut into much more intricate pieces than physical ones: details like atomic structure tend to get in the way in the real world.

The second major problem is that the shapes you have to cut the pieces into are so bizarre, they don't have well-defined volumes.

The Axiom of Choice allows you to cut up a ball and rebuild it with a twin.

If you assume the Axiom of Choice to be true, then the Banach-Tarski Paradox works in three or more dimensions, but not in one or two — three-dimensional space (and especially three-dimensional rotations) is much more complicated than two-dimensions.

However, in two dimensions, you have the von Neumann paradox: you can split a square up into pieces, apply transformations that don't change their area, and reassemble them to make two squares the same size as the first.

One of the most famous advanced math jokes goes: did you know "Banach-Tarski" is an anagram of "Banach-Tarski Banach-Tarski"?

I don't know about you, but the first time I heard it, I was beside myself.

THE AXIOM OF CHOICE

The idea of cutting things into pieces and ending up with twice what you started with depends, as previously mentioned, on the Axiom of Choice — so just what is the Axiom of Choice?

Well, it's the fairly reasonable idea that the set you get from combining two nonempty sets is, itself, not empty.

Informally, if you've got a number of bags — even an infinite number of bags — that each contain balls, you can make a selection of exactly one ball from each bag.

If the number of bags is finite, or there's something that distinguishes the balls, there's no problem. However, with an infinite number of bags and identical balls, there's nothing in the other axioms of set theory to say you can do this. In fact, Kurt Gödel proved that the opposite axiom was consistent with the other axioms — meaning you have a free choice about whether to accept the Axiom of Choice.

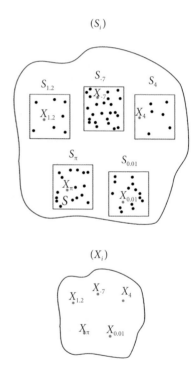

If the elements in each of the sets in S_i are mutually exclusive then it follows that there must exist a set X_i containing one element from each of the S_i sets.

Russell imagined infinite numbers of socks and shoes when considering the Axiom of Choice.

Although it's used uncontroversially by most set theorists, Tarski's original paper connecting the *cardinality* (size) of two related infinite sets was rejected by two referees: one, Fréchet, because the two things he was relating were obviously true, so it didn't count as a new result; the other, Lebesgue, because the two things he was relating were obviously false, so the work was of no interest.

The Axiom of Choice is necessary to select a set from an infinite number of socks, but not an infinite number of shoes.

Bertrand Russell

CHAPTER 12
PLAYING GAMES

One man decides how to make decisions, TV game show contestants play a prison game, computers learn to have fun, a computer geek gambles his way to a fortune and we open a door to win a Cadillac.

JOHN VON NEUMANN

Like Alan Turing, John von Neumann (1903–1957) would fit perfectly well in several of the chapters of this book. He had a finger in pretty much every interesting scientific pie in the first half of the 20th century.

Quantum physics? Yes, he introduced operator notation. DNA? Yes, he predicted self-replication before it was discovered. The Manhattan Project? Yes, he was there. The question of undecidability? Yes, he introduced an important set theory axiom. Fractals? Yes, he introduced continuous geometry, which allows fractional dimensions. Monte Carlo simulations? Yep, pretty

much invented those. Measure theory. Ergodic theory. Operator theory. Lattice theory. Quantum logic. Mathematical economics. Linear programing. Statistics. Fluid dynamics. Computing …

By the time von Neumann was my age, he'd made significant contributions in every topic on that list — many of which I've barely heard of. Irritatingly, none of those are even his most significant contribution to math.

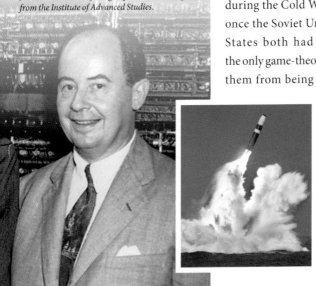

John von Neumann, far right, poses with colleagues from the Institute of Advanced Studies.

Von Neumann, originally from Hungary, invented game theory, the study of what the best decision to make is. In 1928, he proved the minimax theorem for two-player games with perfect information. The best strategy in such a game is the one where the worst-case scenario is the least awful. He later extended this to cover games of imperfect information, and games with several players.

One place where von Neumann famously applied game theory was during the Cold War. He realized that, once the Soviet Union and the United States both had nuclear weapons, the only game-theoretical way to prevent them from being used was to ensure that retaliation would completely obliterate the other side. Neither side would have the incentive to start a war, or to disarm.

Could game theory help to keep the peace?

Neumann called this doctrine Mutually Assured Destruction — or MAD for short. There was a teensy problem with it — it relied on humans not making mistakes or getting cross, which is as crazy as it sounds. Several times in the 40-odd years when the USA and the USSR were at loggerheads in the Cold War, nuclear Armageddon was avoided more by luck than judgment.

Von Neumann was renowned as a genius — his memory was phenomenal, his mental arithmetic spectacular, and his speed of thought unparalleled. Isaac Halperin said that trying to keep up with him was like chasing a racing car on a tricycle.

THE PRISONER'S DILEMMA

There was once a game show on British television called *Goldenballs*, a cut-throat game of bluff and back-stabbing, hosted by Jasper Carrott, a comedian best-known for making snide remarks about foreign commercials. *Goldenballs*, though, was compelling TV.

After a few rounds of skulduggery, all but two contestants were eliminated, and — as game shows like to do — they were pitted against each other to play or a jackpot. Each player was given two golden balls: one with the word "Split," and the other with the word Steal."

If both players chose "Split," it was all sunshine and roses: each would

The Goldenballs *game show involved bluff and double-bluff, and was based on a problem from game theory.*

Both prisoners face a dilemma: should they rat on the other or not?

take home half of the jackpot. If one player chose "Steal" and the other chose "Split," the stealer would take home the entire jackpot, while the splitter would leave with nothing. If both picked "Steal," though, both would leave empty-handed.

Although the thrill of *Goldenballs* was watching two nonmathematicians trying to convince each other that they were lovely people and not about to do something as devious and nasty as try to steal the jackpot before doing exactly that, the final round was based on a famous problem of game theory: the prisoner's dilemma, first posed by Merrill Flood and Melvin Dresher in 1950.

The traditional set-up involves two members of a criminal gang, both of whom are arrested on flimsy evidence.

Each is offered a choice: they may exercise their right to remain silent, or they may incriminate the other. If both remain silent, they will both be convicted of a minor charge, and probably get a couple of years in prison.

If one prisoner rats and the other stays silent, the informant will walk free, whereas the silent party can expect a long prison sentence. If, on the other hand, each criminal incriminates the other, both will get moderately long sentences. The usual language is to "cooperate" by staying silent or to "defect" by giving up their so-called friend.

It is similar to *Goldenballs* in that the best outcome all round is for both prisoners to cooperate — but in both cases, there's a twist.

The longest sentence is reserved for the prisoner who cooperates while their opponent defects.

If you were playing either game, your best option is to defect. If the other person chose to cooperate (or split), you would get a lighter sentence (or a bigger jackpot) by defecting (or stealing).

If they chose to defect, you would get a moderate sentence by defecting, or a very long sentence by remaining silent — so you'd do better to defect in that case, too. In *Goldenballs*, if your opponent steals, you're getting nothing in either case; but if you steal, too, you get to watch the grin on their face vanish, which is worth any number of jackpots, if you ask me.

The game changes if you extend it to multiple rounds when a strategy of tit-for-tat, or doing what the other player did last time, tends to outperform simple defection — but that's a story for another book.

Meanwhile, the prisoner's dilemma isn't simply a toy problem for use in game shows and what-ifs, it can be used to model many situations in which competition and cooperation are options — from doping in sport, to spending on advertising, to freeloading, to nuclear arsenals.

The shortest combined sentence comes from cooperation, but can you take the risk?

Humans can still defeat computers in the Chinese game of strategy, Go, which has an enormous set of possible positions.

PLAYING SERIOUS GAMES

Programing a computer to play Xs and Os isn't an excessively hard problem. There are few enough moves and solutions to the game for a computer to run through them in the blink of an eye.

Even if you ignore symmetry and games that finish before all nine squares are filled, there are fewer than 400,000 possible arrangements of Xs and Os on a three-by-three grid. Other simple games — like Nim or nine men's morris — are similarly straightforward: the number of possible situations in each game is (for a computer, at least) small.

At the other end of the spectrum, there are two giants: the game of Go, which has something on the order of 10^{170} legal game situations, and chess, which has about 10^{50}. You can compare this to moderately complex games like checkers — 10^{18} or so — and Connect 4 — maybe 10^{10}, and realize that there's a wide, wide gulf between them.

The first serious attempt at having a computer play a nontrivial game was Arthur Samuel's work on checkers in the late 1950s. Rather than try to evaluate every position to the end of the game, Samuel developed a system for analyzing the strength of a position: who has more pieces? Who has more kings? Who has more pieces in strong positions? By following every possible move for a few iterations, his program could then apply the minimax strategy to pick a move. It played at a good amateur level, considered tricky but beatable by opponents at that level. Its middle-game was excellent, but it was poor in its openings and endgame.

Checkers is, to the best of my knowledge, the most complicated game so far to be completely solved — with perfect play by both players, it ends in a draw.

This was the result of two decades of work by the team who developed Chinook, the first computer program to beat Marion Tinsley, widely thought to be the greatest human player of all time. In a 45-year career, depending on how you count, he lost only seven games ... two of them were to Chinook.

Draughts is a moderately complex game that has now been completely solved by computers.

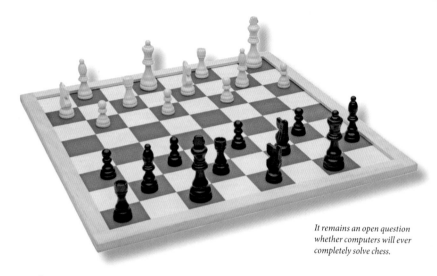

It remains an open question whether computers will ever completely solve chess.

Chess, on the other hand, is still a long way from being solved. I remember watching Deep Blue narrowly losing to world champion Garry Kasparov in 1996 before defeating him the following year.

Kasparov accused the IBM team behind the software of cheating — although there's a good chance that the "unusually creative" move it made in the match that it won was the result of a bug. Today's chess computers are much faster, much more efficient, and have much deeper databases than Deep Blue, and even on modest hardware, humans are no longer a match for our silicon overlords.

As for Go, even the best computer players require significant handicaps to overcome the top human players, although the gap is closing consistently.

Norwegian Magnus Carlsen became a chess grandmaster aged 13.

ELO RATINGS

The standard way to compare the relative strengths of two chess players (or, in principle, players or teams in any sport) is to use the Elo system. It's a predictive system based on the normal distribution, although there is some dispute about whether this is the most accurate distribution to use. Basically, you would expect a player with an Elo lead of 100 points over an opponent to win around two-thirds of the matches, rising to three-quarters for a lead of 200 points.

It's a complicated system. After each game (in reality, at the end of each month), points are redistributed from losing players to winning ones: if you beat someone much inferior to you, you can expect only a few points, but if you upset a grandmaster, your rating will skyrocket!

An average club chess player will have an Elo rating of about 1,500; the top-ranked player in the world (at the time of writing) is the Norwegian Magnus Carlsen, who has held the highest-ever human rating, just short of 2,900.

By contrast, the current leading computer programs have ratings in the region of 3,300 — meaning the ratings system predicts they would take 91 points out of 100 against Carlsen.

In reality, of course, Carlsen would most likely adjust his tactics significantly to find ways of avoiding defeat.

Now the world number one, Magnus Carlsen is still predicted to lose against the best computers.

CHRIS "JESUS" FERGUSON

The world of poker is murky, to say the least. In its spiritual home, the U.S., casual games are effectively illegal in many states, which leads to shady characters meeting in smoky back rooms.

It's a world of names like Amarillo Slim and Annie Duke, a world of graduates of the school of hard knocks, a world of arcane language and superstitions. It's not the kind of world where you'd expect a fresh-faced computer science student to walk in and start a revolution, but that's exactly what Chris Ferguson did in the late 1990s.

Son of a professor of game theory, Ferguson started playing poker aged 10, and began applying the ideas of game theory as soon as he could, learning to exploit mathematical weaknesses in other people's play — to the point where, in 2000, he beat old hand T. J. Cloutier to win the main event at the World Series of Poker.

Before Ferguson, poker was largely a game of feel and guts. There was math involved, certainly: you could make

Chris "Jesus" Ferguson applies the lessons of game theory to work out the best poker strategies.

Mathematics teaches poker players that betting more aggressively than your opponents is often the winning strategy.

rough estimates of what to do based on how many cards were left in the deck, and make educated guesses of how likely you were to win depending on your opponents' actions over the course of the game.

It turned out that the game had been played rather too passively for a long time, which Ferguson could make the most of by upping his level of

"aggression" — not physical aggression, he's renowned for sitting impassively under a wide-brimmed hat — but the amount he was willing to bet.

The next time you're tempted to ask, "When am I going to use probability in real life," have a little think about Chris Ferguson, and notice that he has won more than $8,000,000 from applying it to a card game.

LET'S MAKE A DEAL

At the start, you have a ⅔ chance of choosing a door with a goat behind it.

It's become a legendary probability problem, and now shows up in high school discussions of probability, but it was relatively unheard-of until *Parade* magazine columnist Marilyn vos Savant wrote about the Monty Hall problem in 1990. If there had been an internet to speak of in those dark days, it would have gone crazy; as it was, the *Parade* mailbag overflowed.

In the final round of the game show Let's Make A Deal, *host Monty Hall would offer you a choice of three doors. Behind one of the doors? The star prize of a brand new Cadillac. Behind the other two? Smelly old goats. You pick a door.*

M onty Hall, who knows exactly where the car is, opens another door to reveal that there is a goat behind it, and offers you the choice of whether to stick with your original choice, or to switch to the third so-far unopened door.

You double your chances of winning the car from ⅓ to ⅔ if you switch after Monty has revealed a goat behind one of the other doors.

Vos Savant explained, patiently, that switching was generally beneficial. My best argument is: if you were originally right, you win the car if you stick. If you were originally wrong, you win the car if you switch. Since you were more likely to be wrong than you were to be right to start with, you will win by switching more often than you will by sticking.

The readers of *Parade* disagreed with this analysis. They received somewhere around 10,000 letters, including around 1,000 from people boasting PhDs, most asserting that vos Savant was an irresponsible idiot propagating lies and errors. Even Paul Erdős, one of the greatest mathematicians of the 20th century, refused to accept that switching was better until shown a solution.

A mischievous variant of the situation is known as the Monty Hell problem: in this case, the host *doesn't* know

In the original game show, the host Monty Hall had control over the way the game progressed.

Thousands of correspondents disagreed with vos Savant, including mathematics professors. They were wrong.

where the car is, but the rules are otherwise the same. If this Monty opened a door you didn't pick to reveal a goat, should you still switch?

In this case, the answer is, "It doesn't really matter." The difference is that, because Monty isn't constrained to opening a goaty door, when he reveals a goat, that provides some rather weak evidence that your original guess could be correct. Given that you can see a goat, the probability of you having originally picked the car is now 50 percent — so either strategy is as likely to win.

You can set up further, even more mischievous variations, such as not always allowing Monty to offer the player a switch — for example, an evil Monty may only offer the switch if the player had originally guessed right (in which case, sticking is the correct move), or an angelic Monty may only offer a switch if the player was originally wrong (switching would be a no-brainer).

Like all games, the best strategy in *Let's Make A Deal* depends on what, precisely, the rules are.

CHAPTER 13
CRACKING CODES

Julius Caesar hides some messages, Al-Kindi cracks them, codes get more complicated and a genius gets hounded to death.

Date	B 0857	Freq	Link	From P23 To Decode	To End P1 TO P23	Serial No KN/WB 6773
14/2/45	TE 0954	7691	14/2			

"TYPED"

N/A

-|-B++L-|-TAG.DER.UEBERNAHME.DES.RGTS++MN--.(C+L-|-SEIT.WANN|
B)

D. HUT 3
14/2
T.E. 0954
F. 7691
M.K. 14/2
L. Whiting
No. WB 6773

|ALS.RGTS++M==.FUEHR++M--|IM.KAMPFEINSATZ.++.VV--.OKH++X--PA.AG

.P.++/QXR.---.ABT++M-|++K--Z++L--.I|+M--A++M--.GEZ++M--|SCHN
1/u

IEWIND++N--.OBERST.U++M--.ABT++M--.CHEF.++Z--.SASASASASA...+

D. HUT 3.
14/2
"TYPED" T.E. 0954
F. 7691
M.K. 14/2
L. Whiting
No. WB 6773

+Z--.---.HOKW.++.QPUWQ.QEMWM.QPPP.K--HZPH++X--FF++Q.QYMOUL.VV
10721 13/2 1000 (/ 116897)

--.AN.H++M--.GR++M--.KURLAND++X+-STOHI++M.VV--.BETR++G+-.BET

REUUNG++MA.QML+.GEN.D++M--.FREIW++M--.VERBAENDE.IN.++.QT--|
15

BITTET.UM.MITTEILUNG++N--.WELCHE.NICHTOSTVOELKISCHENFREIW++A|

--VERBAENDE++N--.GRUPPEN.U++M--.EINZELFREIW++MN--.DIE.NICHT.|

EARLY CIPHERS

For almost as long as humans have communicated with each other, they've wanted to do so in secret.

There are boring ways to do that — you could memorize the message and go to a private room, for example, or you could conceal your communication somewhere about the person of a messenger (which is where the phrase "keep it under your hat" is supposed to come from). You might write a normal letter and put pin-pricks under letters spelling out your secret message. A more modern technique, steganography, involves hiding the contents of a message in, for example, a digital picture by subtly adjusting the colors in some part of the image — changes not visible to the human eye, but easily extracted by computer.

All of these are terribly clever, but not necessarily mathematical, ideas. As far as we're concerned, the math of codes begins with the substitution cipher.

The ancient Greeks used written codes. This image painted on a vase shows a student writing on a tablet, although it looks more like a laptop.

KEY TERMS

Plain-text: the message you're trying to hide.
Cipher-text: the encrypted message you submit.

This is a very simple idea: you replace each letter with another, fixed letter. You might replace every A in your plain-text with a V, every B with a P, and so on. When your message arrives, your correspondent runs the map backwards to find your message. You might also divide the enciphered message into equal bites.

A 3-STEP SUBSTITUTION CIPHER

Plain-text alphabet: ABCDEFGHIJKLMNOPQRSTUVWXYZ
Cipher-text alphabet: ZEBRASCDFGHIJKLMNOPQTUVWXY

1) A message of

TAKE CARE HE IS WATCHING

2) encipher

QZHA BZOA DA FP VZQBDFKC

3) equal bites

QZHAB ZOADA FPVZQ BDFKC

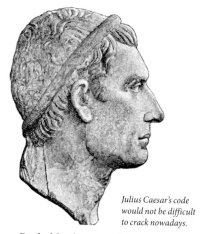

Julius Caesar's code would not be difficult to crack nowadays.

X would become A, Y goes to B and Z goes to C. Decoding a Caesar cipher is as simple as moving each letter back the right number of spaces.

Another example is found in Edgar Allan Poe's short story *The Gold-Bug*, in which the main characters hunt for (and find) Captain Kidd's buried treasure by decoding a cipher using the technique of frequency analysis.

Probably the most famous example of a substitution cipher is the Caesar, or shift, cipher, in which each letter moves forwards or backwards a certain number of letters in the alphabet — so, using a shift of three, A would become D, B would become E, C would become F and so on; it would "wrap around" at the end of the alphabet, so W would become Z,

This is the problem with using a substitution cipher: they're very easy to crack once you know what you're doing — Caesar probably got away with it because not many people could read Latin at the time, let alone Latin in code. Luckily for the more secretive among us, there are more secure methods to hide messages that should remain hidden.

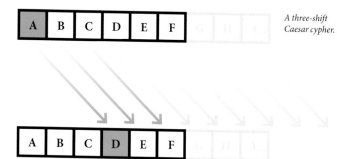

A three-shift Caesar cypher.

If you've ever played Scrabble, or Words With Friends, or similar word games, you'll know that some tiles — Q, Z, J and K, for example, are worth more points than others, like E, T and A. Why are these letters worth more? It's because they're harder to use. Why are they harder to use? They're in fewer words.

Once you have worked out that some letters inevitably appear quite frequently whereas others are hardly used at all you can apply that knowledge to a cipher and experiment with substituting letters.

Captain Kidd welcomes guests aboard his ship in New York Harbor. Treasure hunters in Edgar Allan Poe's The Gold-Bug *searched for his loot by decoding a cipher using frequency analysis.*

CRACKING CAESAR (AND OTHER SUBSTITUTION) CIPHERS

If you know the language in which a code is written — English, Spanish, French or German, for example — and you have studied the language to establish how frequently each letter or symbol is used, you are well on the way to cracking the code.

If you take a random, longish block of English text and count how often each letter shows up, you'd expect to see a letter E, the most common letter, about one in every eight letters. T is the next most common (about one in 11) followed by A (about 1/12), O (1/13), I (1/14) and N (1/15). The rarest letters are Q (1/1,050), Z (1/1,350), X and J (both about 1/700).

This presents a major problem for a substitution cipher that wants to keep its secrets. By counting the number of times each letter shows up in the

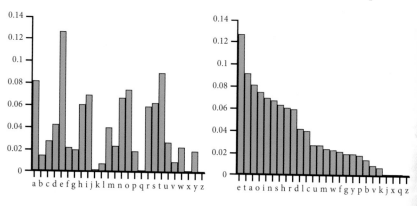

These graphs show how frequently letters are used in the English language.

cipher-text, you can make an educated guess at what it must represent. For instance, if your cipher-text has 200 characters, of which 25 are Vs, you might well hypothesise that a cipher-text V stands for a plain text E. You'd try replacing each V with an E, and see if any other words started to become clear.

There are other weaknesses, too. Some letters are much more common as doubles than others (you see EE and LL all over the place in English, but, unless you're a glowworm, WW almost never shows up). Others occur most commonly in certain combinations — Q is almost always followed by a U. Add to this that you can often spot words (if you have T?E showing up in several places, you can have a pretty fair guess that the missing letter is H — unless the person you're trying to eavesdrop on is discussing neckwear).

Sadly for the eavesdroppers among us, the substitution cipher is considered a bit of a baby code these days — you might find it as a newspaper puzzle, but not for serious codes.

Frequency analysis helps to explain the different values of letters used in Scrabble.

AL-KINDI

Abu Yūsuf Yaʻqūb ibn ʼIshāq as-Sabbāh al-Kindī (c. 801–873CE) was considered by Cardano to be one of the 12 greatest minds of the Middle Ages.

He is said to have written more than 260 books, including 32 on geometry and 12 on physics — but few, alas, have survived. He was also an influential writer on philosophy, theology, medicine and music, he wrote a series on books explaining how the new-fangled Indian number system — which we now call Arabic numerals — worked, and tried to explain that infinity was a very bad idea.

Born in Kufa to a line of governors, al-Kindi eventually moved to Baghdad to study and won the patronage of the Caliph

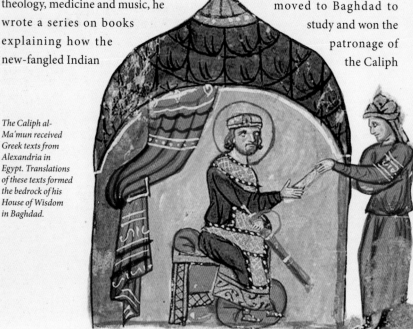

The Caliph al-Ma'mun received Greek texts from Alexandria in Egypt. Translations of these texts formed the bedrock of his House of Wisdom in Baghdad.

al-Ma'mun, who was in the process of creating the House of Wisdom, a proto-university in which the main goal was to translate the knowledge of the Greeks into Arabic. Like all good scientists, he took the existing literature and developed it further.

Unfortunately, the politics of the Caliphate eventually worked against al-Kindi; whether for religious or academic reasons, he fell out of favor with the rulers, and he is said to have died a lonely man.

My main reason for mentioning him is this: he is generally credited as the inventor of frequency analysis. Since cryptography is, by necessity, a secretive sort of endeavor, it's especially unusual to be able to say "this person was responsible for this technique."

POLYALPHABETIC CIPHERS

As soon as people with secrets realized that their secrets weren't as secret as they had hoped, they started work (in secret, of course) on ways of making their secrets even more secret.

The next step up from a simple substitution cipher is a *polyalphabetic* cipher — one that uses multiple substitution alphabets. Although it's thought that al-Kindi may have invented such a cipher in the 9th century, the first documented example is credited to Italian Leon Battisti Alberti in 1467. The Alberti cipher involved sticking with a substitution cipher for as long as he fancied, then switching to another, which he indicated by using a capital letter or a number.

A few decades later, in 1499, Johannes Trithemus wrote the *Steganographia*,

Poet, artist, architect, priest and cryptographer, Alberti was a true Renaissance man.

which wasn't published until 1606 and was almost immediately banned. The book is, on the face of it, about using spirits to communicate over long distances — but once the content is deciphered, it turns out to be about cryptography (the Catholic Church eventually removed it from the *Index Librorum Prohibitorum* in 1900). One of his codes involves changing the alphabet used to encode the text after every letter: you might encode the first letter of your text using a Caesar shift of one, then the second letter by a shift of two, and so on.

In his 1467 treatise De Cifris, Alberti describes a cipher disk made of two concentric rings that rotate independently. The message to be encrypted was spelt out on the outer ring, and encoded onto the inner ring, which could be spun to a different position at any time.

Unfortunately for 15th-century secrets, both Trithemus's and Alberti's codes are straightforward to break. Luckily for 16th-century secrets, the Italian Giovan Battista Bellaso had a better idea (a very similar idea appeared 30 years later by Frenchman Blaise de Vigenère, and it's typically called a Vigenère cipher).

Instead of using a small set of alphabets at random, or a predictable set like Trithemus, Bellaso decided on using a keyword or phrase. For example, if he picked the key phrase SECRETPROJECT, the first letter would be encoded using a Caesar shift of 18 (moving A to S, the first letter of SECRETPROJECT), the second using a shift of 4 (moving A to E, the second letter of the key phrase) and so on. When you reach the end of the key phrase, you go back to the start: SECRETPROJECT has 13 letters, so the 13th letter of your code would be shifted by 19 (A to T, the last letter of the key phrase) and the 14th by 18 (A to S again).

German abbot Johannes Trithemius dabbled in the occult as well as secret codes.

This was a significant improvement — unless you knew what the key phrase was, you were a bit snookered! The cipher was so notoriously difficult to break that it was known in French as *le chiffre indéchiffrable*, the undecipherable cipher. It wasn't until 1863 that Kasiski published a method for breaking it.

And that, you would think, was that ... at least until the Enigma came along.

Plain-text: **ATTACK AT DAWN**
Key: **SECRETPROJECT**
Cipher: **SXVIGDPKRJAP**

	A	B	C	D	E	F	G	H	I	J	K	L	M	N	O	P	Q	R	S	T	U	V	W	X	Y	Z
A	A	B	C	D	E	F	G	H	I	J	K	L	M	N	O	P	Q	R	S	T	U	V	W	X	Y	Z
B	B	C	D	E	F	G	H	I	J	K	L	M	N	O	P	Q	R	S	T	U	V	W	X	Y	Z	A
C	C	D	E	F	G	H	I	J	K	L	M	N	O	P	Q	R	S	T	U	V	W	X	Y	Z	A	B
D	D	E	F	G	H	I	J	K	L	M	N	O	P	Q	R	S	T	U	V	W	X	Y	Z	A	B	C
E	E	F	G	H	I	J	K	L	M	N	O	P	Q	R	S	T	U	V	W	X	Y	Z	A	B	C	D
F	F	G	H	I	J	K	L	M	N	O	P	Q	R	S	T	U	V	W	X	Y	Z	A	B	C	D	E
G	G	H	I	J	K	L	M	N	O	P	Q	R	S	T	U	V	W	X	Y	Z	A	B	C	D	E	F
H	H	I	J	K	L	M	N	O	P	Q	R	S	T	U	V	W	X	Y	Z	A	B	C	D	E	F	G
I	I	J	K	L	M	N	O	P	Q	R	S	T	U	V	W	X	Y	Z	A	B	C	D	E	F	G	H
J	J	K	L	M	N	O	P	Q	R	S	T	U	V	W	X	Y	Z	A	B	C	D	E	F	G	H	I
K	K	L	M	N	O	P	Q	R	S	T	U	V	W	X	Y	Z	A	B	C	D	E	F	G	H	I	J
L	L	M	N	O	P	Q	R	S	T	U	V	W	X	Y	Z	A	B	C	D	E	F	G	H	I	J	K
M	M	N	O	P	Q	R	S	T	U	V	W	X	Y	Z	A	B	C	D	E	F	G	H	I	J	K	L
N	N	O	P	Q	R	S	T	U	V	W	X	Y	Z	A	B	C	D	E	F	G	H	I	J	K	L	M
O	O	P	Q	R	S	T	U	V	W	X	Y	Z	A	B	C	D	E	F	G	H	I	J	K	L	M	N
P	P	Q	R	S	T	U	V	W	X	Y	Z	A	B	C	D	E	F	G	H	I	J	K	L	M	N	O
Q	Q	R	S	T	U	V	W	X	Y	Z	A	B	C	D	E	F	G	H	I	J	K	L	M	N	O	P
R	R	S	T	U	V	W	X	Y	Z	A	B	C	D	E	F	G	H	I	J	K	L	M	N	O	P	Q
S	S	T	U	V	W	X	Y	Z	A	B	C	D	E	F	G	H	I	J	K	L	M	N	O	P	Q	R
T	T	U	V	W	X	Y	Z	A	B	C	D	E	F	G	H	I	J	K	L	M	N	O	P	Q	R	S
U	U	V	W	X	Y	Z	A	B	C	D	E	F	G	H	I	J	K	L	M	N	O	P	Q	R	S	T
V	V	W	X	Y	Z	A	B	C	D	E	F	G	H	I	J	K	L	M	N	O	P	Q	R	S	T	U
W	W	X	Y	Z	A	B	C	D	E	F	G	H	I	J	K	L	M	N	O	P	Q	R	S	T	U	V
X	X	Y	Z	A	B	C	D	E	F	G	H	I	J	K	L	M	N	O	P	Q	R	S	T	U	V	W
Y	Y	Z	A	B	C	D	E	F	G	H	I	J	K	L	M	N	O	P	Q	R	S	T	U	V	W	X
Z	Z	A	B	C	D	E	F	G	H	I	J	K	L	M	N	O	P	Q	R	S	T	U	V	W	X	Y

In Trithemius's tabula recta, each row is made by shifting the previous one to the left.

KASISKI EXAMINATION

Kasiski's breakthrough in cracking polyalphabetic ciphers was a two-step process: first of all, work out how long the code phrase is; and once you know that, split the cipher text up into rows of that length and decode each column using frequency analysis.

The second part is easy; it's the first that's tricky. Kasiski's method was to look for strings of characters repeated in the cipher text — ideally, three or more characters long. He would then find the distance (in characters) between the starts of these groups. Barring the odd coincidental repetition, the distances must be multiples of the key length.

That's a bit tedious to do by hand, but much simpler by computer. However, it can be made even more efficient by a method called superposition. You place

Friedrich Wilhelm Kasiski was a German infantry officer and codebreaker.

two copies of the cipher text side-by-side, and offset one of the copies first by one character, then by two, and so on. In each case, you count how many times the same character appears in the same place in both the fixed text and in the offset text. When the offset is a multiple of the key length, the number of coincidences tends to increase markedly.

KCDVR	ZWEXC	NSEDM	JSSZX	SFIVY	FRZEC	PDCZA	SPCVR	ZSIVG	KOEFR	ZSIKF	WCIPU	ZWTYO	LOKVQ	LVRKR
HDVRP	SBUSC	JSGCY	USUSW	KCDVR	ZWEXC	NSEDM	JSSZX	SFIVY	FRZEC	PDCZA	SPCVR	ZSIVG	KOEFR	ZSIKF

Kasiski placed copies of the cipher text next to one another, offset one of them and counted how many characters appeared in the same place. When the offset is a multiple of the key length, the number of coincidences rises.

CIPHER INTERCEPT

Cipher intercept:

OSRGM DCXZQ WTFIR ZSZEA GBMVL ASETC

Keyword:

SORRY

Plain-text:

WEAPO LOGIS EFORT HEINC ONVEN IENCE

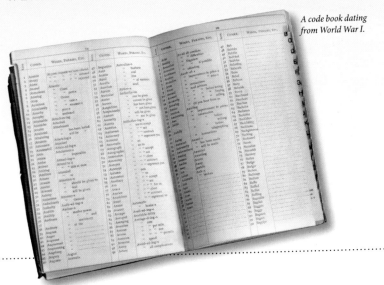

A code book dating from World War I.

BLETCHLEY PARK

In January 1945, as World War II was drawing to a close, 9,000 people worked at a single site on a single project. In the early 1970s — apart from the people directly involved — there was still practically nobody who knew about the work, or its importance.

The site, a stately home, conveniently located close to England's main north-south and east-west railway lines, was Bletchley Park, home of the British Government Code and Cipher School. At its heart was an oddball team of boffins (it's claimed by some that the word was invented to confuse German spies): linguists and crossword fiends, engineers and mathematicians — most famously, Alan Turing. And the "professor-type" was needed: the Axis codes they were trying to break — Enigma and Lorenz — were practically unbreakable, if used properly.

An Enigma machine consists of a keyboard connected to a series of rotors, connected to a switchboard. When you press a key, a current passes through the rotors (each rotor implements a different substitution cipher), through the switchboard (yet another cipher) and back, eventually turning on a light to show you the encoded message. The rotors then rotate, meaning the next key will be encrypted by a totally different

cipher. Effectively, a three-rotor Enigma machine was a polyalphabetic cipher with a keyword length of nearly 20,000 — and as it was used for short messages, Kasiski examination techniques would be no use at all. The number of possible keywords was almost 1.6×10^{20}. To put that in context: if you tried 400 different settings every second, it would take you pretty much the entire age of the universe to check them all.

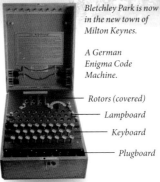

Bletchley Park is now in the new town of Milton Keynes.

A German Enigma Code Machine.

Rotors (covered)
Lampboard
Keyboard
Plugboard

Much of the work in understanding how the Enigma machine worked had been undertaken in Poland before the war began — Marian Rejewski determined the wiring in the Enigma rotors using permutation theory — an extension of Galois's work a century before. Rejewski also spotted a flaw in how the machines were used: a single base setting was used for all messages on any given day, which was used to encrypt whichever three-letter setting the operator felt like using. That would have been ok — only the operators would then repeat the key, just to be sure.

That opened the whole thing up. Knowing that the first letter and the fourth letter were connected allowed Rejewski to find patterns — and the patterns dramatically reduced the number of possible rotors, down from hundreds of quintillions to the order of 100,000, which is the work of hours to solve.

Or less, if you have a machine. Rejewski was the first to come up with the idea of a *bombe*, a simulated

Enigma which would span through all of the possible combinations until it hit the one that must have been used. It was this idea that Turing and Harold "Doc" Keen built on at Bletchley to develop a more sophisticated bombe of their own — one you can still see operating if you visit the museum at Bletchley Park today.

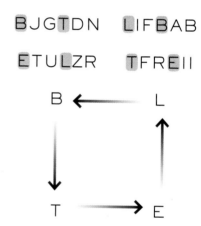

A four-cycle formed from the first and fourth letters of four Enigma message indicators, as used before 1938 by the German military.

Turing and Keen's bombe has been reconstructed at the Bletchley Park Museum. Each of its rotating drums simulates the action of one of the rotors on an Enigma machine.

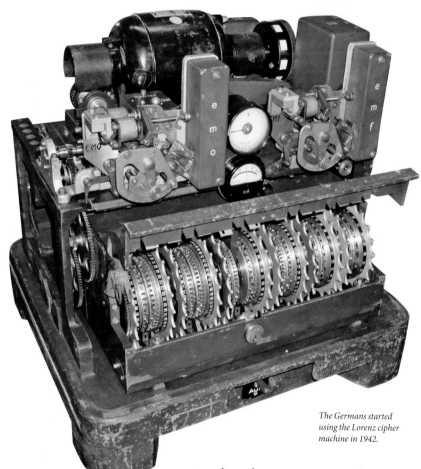

The Germans started using the Lorenz cipher machine in 1942.

The Lorenz cipher was even tougher than the Enigma machine. It required a lucky break (a retransmitted, slightly altered, message using the same settings) and a stroke of genius from cryptanalyst Bill Tutte, who deduced the entire structure of the system from roughly 4,000 cipher text characters.

His team at Bletchley built a machine to decipher Lorenz that was completely unrelated to the machine that had encrypted the messages, because they'd never seen one. This achievement is thought of as one of the greatest intellectual feats of the entire war, and led to the building of the Colossus, possibly the world's first programmable computer, to determine the necessary settings.

With an enormous combination of intelligent guesswork, mathematical analysis, resilience to failure, and mechanical (not to mention human) power, the British were able to intercept and decipher a large portion of German radio traffic — intelligence which, historians believe, shortened the war by at least 2 years.

My favorite Bletchley Park story comes from when the work was finally declassified in the 1970s. A husband sat his wife down and said "Darling — I have something to tell you: I worked at Bletchley Park during the war." The wife replied, "Really, my dear? So did I!"

> [The Bletchley Park Staff were] *the geese that laid the golden eggs and never cackled.*
>
> Winston Churchill

The 1944 D-day landings depended on information from code-breakers.

ALAN TURING

In terms of influence on modern life, it's hard to think of a more significant mathematician than Alan Turing (1914–1952).

Before World War II, he (along with Church) cracked the *Entscheidungsproblem*, a significant open problem at the time — and in the process, came up with a specification for a computer. Although there are many, many different computer languages in use today, they can all (theoretically) be reduced to instructions mathematically equivalent to the simple set introduced in Turing's work.

During World War II, he worked on breaking the Enigma and Lorenz codes at Bletchley Park, thinking all the while about whether a computer could think. He came up with what is now known as the Turing Test for artificial intelligence — roughly speaking, if a computer can fool you into thinking it's human, then you can call it intelligent. Working with Jack Good, he also concocted important statistical rules concerning sizes of unobserved populations, and — after the war — began working as a biologist.

A statue of Turing in Manchester shows him holding an apple. An encrypted message decodes to read "Founder of Computer Science."

He was a champion long-distance runner, and would frequently run to conferences rather than take the train. He invented the game of Run-Around Chess, in which you make a move on the chessboard and then run around the house — if you get back to the board before your opponent, he or she misses a turn. It's one of the few sports that combines mental and physical exertion.

Turing had a tragic end. In the 1950s, homosexuality was illegal in Great Britain. He was convicted under these laws and forced to undergo hormone treatment under threat of imprisonment. This could have contributed to his early death: he died after working with cyanide in the biology lab and eating an apple contaminated with the poison. Turing was posthumously pardoned in 2013.

As a supremely fit runner as well as a math genius, Turing had a bit of an unfair advantage in Run-Around Chess.

THE TURING TEST

On Turing's 100th birthday in 2012, his Manchester statue was festooned with flowers and a party hat. I like to think he'd have approved.

It was while working at the University of Manchester that Turing first proposed "The Turing Test," an early attempt to define computer intelligence. Turing suggested that a computer able to fool a panel of human judges into believing it was human could, for all practical purposes, be considered intelligent. In 2014, a computer "passed" by pretending to be a 13-year-old Ukrainian with limited English.

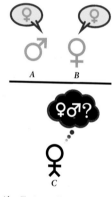

Alan Turing outlined an "Imitation Game" where player C has to decide whether player A or player B is a man through a series of written questions and answers. Player A (the man) attempts to fool player C while player B tries to give assistance.

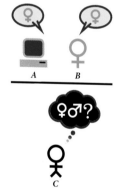

In another version of the "Imitation Game", the role of player A is taken by a computer, while player B continues to try to help player C.

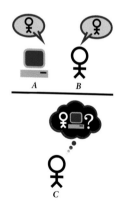

In this version, with player C still issuing written questions and receiving written answers, he must decide which of the others is the human.

Eve wants to eavesdrop

Alice wants to send a message to Bob

Bob receives Alice's message

Peggy wants to prove the message was sent

Victor verifies that the message was correct

Mallory wants to alter the message

MEET ALICE AND BOB

If you read any explanation of a cryptographic technique, the chances are it'll start with something like "Alice and Bob exchange their public keys." No explanation of who these characters are, no back-story, nothing.

Allow me to put that right.

Alice and Bob first showed up in the late 1970s — when Ron Rivest wrote up the RSA protocol, he figured that "Person A" and "Person B" would be difficult to follow, so he opted for personal names instead.

Since then, all manner of conventions have developed regarding the cryptographic *dramatis personae*.

Usually, Alice wants to send a message to Bob. Eve, who is evil, may be attempting to eavesdrop, without altering the message. Mallory, who is also a malefactor, may be trying to launch an active man-in-the-middle attack. If the transaction needs to be checked (of course, without giving away any information), then Peggy the prover and Victor the verifier may be brought into play.

It's one of the most endearing things about cryptography. Where most mathematical problems have technical names that are tricky for the lay person to keep track of, the continuing saga of Alice and Bob immediately allows anyone familiar with them to follow what's going on.

PUBLIC KEY CRYPTOGRAPHY

Clifford Cocks might go down as one of the most unfortunate mathematical discoverers of the 20th century.

Like many others, he missed out on credit for his invention because he didn't publish it. This wasn't due to laziness or ineptitude or any of the other usual reasons for missing the boat. It was because he worked for the British spy agency, GCHQ, and his work from 1973 was only declassified in the late 1990s.

MIT's Stata Center for Computer Science has an appropriately deconstructed modern design.

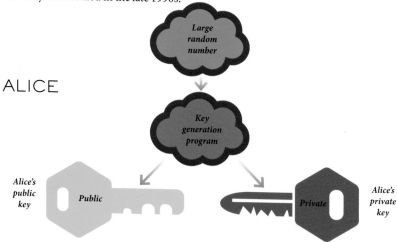

ALICE

Large random number

Key generation program

Alice's public key

Public

Private

Alice's private key

Four years after Cocks did his top-secret work, Ron Rivest, Adi Shamir and Leonard Adelman, three mathematicians at MIT, independently developed an effectively identical system for encrypting and decrypting messages.

It's now called RSA in their honor. It's a remarkable development, no matter who thought of it. The idea is to invite anyone sending you a message to use a public key — one openly accessible to anyone who wants it — to encrypt the message, which you can then decode with a private key that only you know. Here's how it

works, with as much math stripped out as possible: Bob generates a public key, which is a pair of numbers, and a private key, which is a related number. Alice, who wants to send Bob a message, has already been told by Bob what his public key is. She uses the public key to encrypt her message.

If Bob's public key is (n, e), and Alice's message is a number M, then Alice transmits $T = M^e$, worked out modulo n. If his private key is d, he works out T^d, which — because of how he picked the numbers — works out to be M.

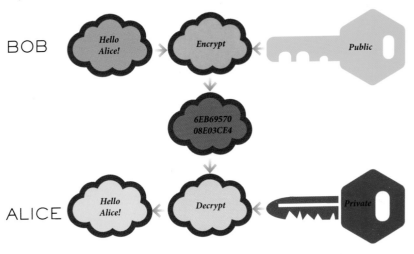

Alice can also sign her message, if she wants to reassure Bob that it's really her sending the message (rather than an impostor). She appends a hash value of her message, encrypted with her private key, to the message. Bob can decrypt this using Alice's public key (the system is symmetrical, so each person's private

GCHQ in Cheltenham, affectionately known as "The Doughnut", is the center for British intelligence gathering and code-breaking.

and public keys unlock each other), apply the hash function to the full message he's received, and check that it matches the signature — if it does, he can be virtually certain the message was sent by Alice, because nobody else knows her private key, and that the message he's received is the one Alice sent, because otherwise the hash value would be different.

The code is designed so that it can only be broken if Eve the eavesdropper can either steal Bob's private key, or figure out how to factorize an enormous number. Factorizing involves breaking a number down into a product of prime numbers. For instance, 15 factorizes into 3 × 5. It's easy for small numbers, but there is no simple solution for large ones. (Recent advances have made factorizing easier, and most public key cryptography now uses elliptic curves instead of factorizing — although the technique is very similar.) As with all codes, it needs to be used properly for it to be secure; with the current state of factorization, if

Bank automated teller machines (ATMs) use RSA encryption to keep your personal details — and wealth — safe from hackers.

your keys are long enough and random enough, your message properly padded and everything else in order, RSA is not thought to be breakable in practice.

CHAPTER 14
A TASTE OF THE 20TH CENTURY

The coastline of Great Britain mysteriously grows every time it's measured, a butterfly fails to cause a hurricane and a strange kind of curve cracks open a 350-year-old conundrum.

Could the flapping of a butterfly's wing cause a hurricane?

BENOÎT B. MANDELBROT

There aren't too many people who invent completely undreamt-of fields of math, but Benoît B. Mandelbrot (1924–2010) was one of them.

Born in Warsaw, Poland, his family moved to Paris in 1936, to join his uncle Szolem. Szolem Mandelbrot was a mathematician who worked at the College de France and inspired his nephew to study math.

When France was invaded by the Germans in 1940, the Mandelbrot family moved out of Paris to Tulle, farther south, where they remained for the duration of France's occupation, forever in fear that the Germans would discover they were Jews.

German soldiers outside the Moulin Rouge, Paris, during World War II.

Mandelbrot returned to Paris in 1944 to continue his studies, also spending time at the Lycee du Parc in Lyon. In 1947, he moved to the United States to study at the California Institute of Technology, where he received his master's degree in aeronautics before returning to France to complete his PhD degree in mathematical sciences at the University of Paris in 1952.

Six years later, he joined IBM in New York, and stayed there for 35 years. During this time, he studied the Julia set, and particularly the Mandelbrot set which bears his name, although he didn't discover it — Fatou and Julia were the first to study it early in the 20th century.

In 1975, he came up with the term "fractal" to describe shapes that display *self-similarity* — the same structures appearing on many different scales. This phenomenon was much more widespread than anyone had expected — from being a mathematical curiosity (one that critics thought was simply an artefact of the way computers

Benoît Mandelbrot gives a speech on being made an officer of the Legion of Honor on September 11, 2006, at the École Polytechnique in Palaiseau near Paris.

handled numbers), it became apparent that fractals abounded in geography (e.g., coastlines), animal biology (e.g., lung structure), plant biology (e.g., broccoli), and finance — the month-to-month movements of the stock market are hard to tell apart from the second-to-second market changes.

Mandelbrot died of cancer in 2010, aged 85.

THE LENGTH OF THE BRITISH COASTLINE

If you have a globe handy, take a look at the island of Great Britain. It's probably toward the top, depending on how you're holding your globe.

If you had to estimate the length of its coastline, you could probably have a rough stab at it. If you know what scale the globe is — you'd measure the coastline (quite roughly, by necessity) and then multiply it by the scale to get an approximate answer.

You'd prefer a better map, though, wouldn't you? So, perhaps you find a map of Europe in an atlas, and do the same thing again, as best you can. You'll get a better, almost certainly larger, estimate for how much seafront Britain has. There's a lot more detail in the atlas — the globe-makers would hardly make a point of including every inlet and headland.

You could do better still,

though: what about getting a poster-sized map of the whole island? Now you can come up with an even better estimate — and it'll be larger still. You can keep going with this: maybe get large-scale rambling maps of each section of the coastline and measure those (an even better, and even larger, estimate). Hypothetically, you could get a trundle-wheel and a great deal of equipment for surviving storms and climbing around rock-faces, and measure the coastline, obviously doing your best to account for tides. In a famous paper in 1967, Mandelbrot determined several things about the British coastline (and, by extension, all coastlines).

Great Britain on a globe.

You can measure more of the coastline on a map of the British Isles than you can on a globe…

… or on a map of Europe.

A small scale map of Norway as part of a European map.

A larger scale version of the map of Norway shows that it is self-similar with the smaller one.

The first was a bit strange: the closer you zoom in, the longer the coastline gets. In some sense, the length of the coastline isn't a well-defined number; even if it was fixed, rather than changing at the whims of tides, waves and erosion, you get significantly different answers with different measuring sticks — and these answers don't have an upper limit!

The second was possibly even stranger: if you trace a section of the Norwegian coast on the map of Europe and a section on the large-scale map, it's very difficult to tell which is which. Coastlines are *self-similar* — at different scales, they look pretty much the same.

THE MANDELBROT SET

- Pick a pair of numbers and call them x and y — which are good, solid mathematical names.
- Work out $x^2 - y^2 + x$ and $2xy + y$, and call these two new numbers x and y.
- Repeat the process until either the numbers stop changing, or they become enormous.

What you've just done is decide whether a point belongs to the Mandelbrot Set — they're the ones that stop changing.

The chances are, the point you picked became enormous. Unless you happened to choose x and y such that $x^2 + y^2$ is less than four (treating x and y as coordinates, all of these points lie inside a circle with a radius of 2 centered at 0,0), it will definitely diverge. Even inside that circle, only about 12 percent of points are in the set.

Obviously, it's extremely tedious to work out this process for every possible point in the circle — especially close to the boundary of the set, you might need to do step two hundreds or thousands of times before you can decide whether or not it converges.

In principle, there's no upper limit to the number of iterations you might need. Luckily, we have machines for that sort of thing nowadays.

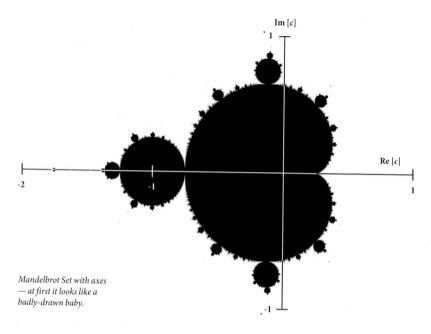

Im [c]
1
Re [c]
-2
-1
1
-1

Mandelbrot Set with axes — at first it looks like a badly-drawn baby.

Computers can quickly and easily do these calculations so that you don't have to — and, more to the point, draw them out nicely.

At first, the Mandelbrot Set looks like a poorly-drawn baby on its side — a large body on the right, a small head on the left and a couple of round arms. However, as you zoom in around the edges, you start to see a much more intricate structure. Smaller babies, bulbs and tendrils; you can zoom in as far as you like, and you'll still see roughly the same shapes.

There are any number of Mandelbrot Set explorers available online — if you get the chance to play with them, do!

The Mandebrot Set isn't just a trippy poster ideal for the average student flat, but a mathematical curiosity in its own right.

For example, if you look at the "pinch points" of the Mandelbrot set — for

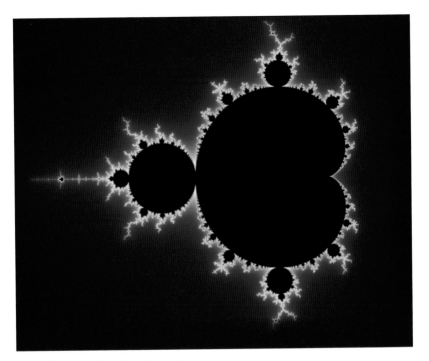

Adding color to a Mandelbrot Set produces striking images.

example, where the body meets the head, or the head meets the hat, places where the set is somehow very thin, you can discover π! If you let $x = -0.75$ and y be a very small number, the number of steps it takes for the process to escape is roughly π divided by your original value of y.

There's also a correspondence between the Mandelbrot Set and the "logistic map," in which the chaotic regions of the map line up with the tendrils of the Mandelbrot Set, and the nonchaotic parts correspond to the bulbs.

FRACTAL LANDSCAPES

Possibly the most common use of fractals outside of pure math is in computer graphics. By taking a shape and recursively altering pieces of it following a recipe involving some random elements, it's possible to generate landscapes almost indistinguishable from reality.

For example, you might take a large, horizontal square, and offset its center vertically by a small, random amount.

You'd then split the square into four smaller squares and repeat the process for each of them. After many iterations, you'll have something that looks like a decent approximation of terrain. More complicated processes, such as multifractals (which take into account that mountains behave differently to plains, and fjords behave differently to beaches, for example), give even more realistic landscapes.

One of the most famous uses of fractal terrain generation is in the science fiction movie *Star Trek II: The Wrath of Khan*, in which an entire alien world is generated algorithmically.

Fractal techniques are also applied in music in the field of algorithmic composition.

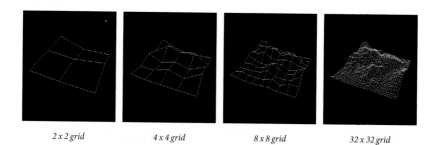

2 x 2 grid   *4 x 4 grid*   *8 x 8 grid*   *32 x 32 grid*

EDWARD LORENZ'S WEATHER SIMULATION

Note: the author has not allowed too many pesky facts to cloud the telling of this story.

Edward Lorenz sat back with a cup of coffee, and smugly put his feet on his desk at the Massachusetts Institute of Technology. His weather simulation was working perfectly: cycles of warm and cold, the occasional rain and sunny spell, the odd snowfall. For a 1960s' computer, this was a magnificent achievement; his smugness was well-deserved.

He took a sip of coffee and screwed up his face. Not because the coffee was bad (although it was), but because he'd realized he had no way of printing out the results of the experiment. He cursed quietly, stopped the program, and printed all of the raw data out on the rickety dot-matrix printer that's probably still sitting on a shelf in an academic cupboard somewhere in Cambridge, Massachusetts.

Lorenz paved the way for the modern weather simulation map.

Lorenz was not expecting his simulation to predict blizzards.

Lorenz sighed, cracked his knuckles, and painstakingly typed the numbers on the printout back into the computer, so it could carry on where it left off, and went to pour himself another cup of coffee.

On his return, he took a sip of his drink and immediately spat it out. Not because the coffee was bad (although it was), but because his simulation had gone crazy. Hurricanes. Droughts. Blizzards. At one point, four horsemen rode across his screen. What on earth could have gone wrong?

He quickly realized that something must have happened when he typed the numbers in — but he'd been careful to check each of them, so he knew there wasn't a mistake there. Unless … the numbers on the printout and the numbers the computer had stored were somehow different?

That was the problem — the numbers he'd typed in from the printer were correct to five decimal places. The numbers stored on the computer, on the other hand, were correct to seven decimal places. But, thought Lorenz, that's a tiny difference. Incredibly tiny. How tiny?

He figured it out: the difference between the numbers on the printout and the numbers in the computer was roughly equivalent to the flapping of a butterfly's wing halfway around the world.

The flapping of a butterfly's wing — comparable to Lorenz's two decimal places oversight.

This is where the legend of butterflies causing hurricanes comes from. It's more complicated than that, of course. The weather is a chaotic system, which means that small changes in initial conditions cause enormous changes in the final outcome.

FEIGENBAUM'S CONSTANT

Here's a little number recipe for you:

1. Pick a number k from the list below (or another value if you prefer).
2. Pick a number between 0 and 1, call it x.
3. Multiply your number by $k \times (1-x)$ and call the result x.
4. Repeat step 3 until you spot a pattern.

Some suggested values of k:

0.5, 1.7, 2.3, 3.2, 3.5 and 3.6

Try others, if you like. You could even make a spreadsheet!

You'll find with small k (less than 1), x gets small very quickly. Middling values of k (between 1 and 3) converge on a fixed value that depends on k. It's between 3 and 4 that things get interesting.

For k between 3 and about 3.49, you end up with x going back and forth between two values. A little higher, and it oscillates between four values. Then eight … until, finally, when k

gets to about 3.57, things go haywire: x bounces around without ever settling down into an obvious pattern.

If you pick two very slightly different values of x and run them through (say) 100 iterations, you end up with two values of x that could be very far apart, or very close together — you simply can't tell. Small variations in initial conditions can lead to large variations in

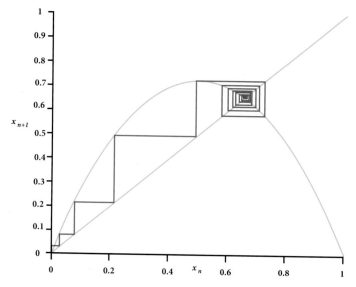

Here, repeated iterations of the function converge on one value, which can be illustrated using a cobweb diagram. The result of the operation is shown on the y-axis. This value is then plotted on the x-axis, and the function is repeated. x_n and x_{n+1} are equal at zero and one other point.

Mitchell Feigenbaum at the Niels Bohr Institute, in Copenhagen, 2006.

the answer — which is the definition of chaos. While investigating the "logistic map," which is what this particular process of messing about with numbers is called, Mitchell Feigenbaum made notes on the values of k where the behavior changed, the "bifurcations." This literally means "splitting into two forks," which is what each solution does.

The bifurcations caused the behaviour to vary from having one stable solution to oscillating between two, from two to four, and so on. He found that the sizes of the gaps between these bifurcations got smaller in a predictable way.

The ratio between one gap and the next turned out to be about 4.669:1. The ratio isn't restricted to this particular process; it shows up in a very broad class of mappings (including the Mandelbrot Set) and is as important in its field as π is in geometry and e in calculus.

This number is now known as Feigenbaum's constant (symbol δ).

$\delta = 4.669\ 201\ 609\ 102\ 990\ 671\ 853\ 203\ 821\ 578$ to 30 decimal places

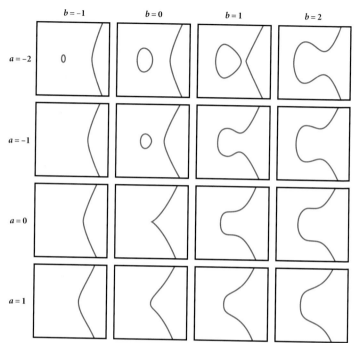

	$b = -1$	$b = 0$	$b = 1$	$b = 2$

Rows labeled: $a = -2$, $a = -1$, $a = 0$, $a = 1$

Elliptic curves are symmetrical around the x axis.

ELLIPTIC CURVES

On the face of it, an elliptic curve doesn't look especially complicated. It's a curve of the form $y^2 = x^3 + ax + b$, where a and b are constants.

It is not, perhaps, the simplest of all curves, but it's lovely and smooth, symmetrical about the *x*-axis. As curves go, it could be a lot uglier.

Sure, for some values of *a* and *b* you get a discontinuous curve (this isn't a problem, as it turns out) and you sometimes need to think about a point at infinity (this is also not a problem, if you have a big head), but compared to, say, fractals, elliptic curves are a model of smoothness and good behavior.

One of their really nice properties involves lines through any pair of points. There are three things that such a line could do:

- It can cross the curve very obediently at a third point.
- If it just grazes the curve at one of the points, it won't meet the curve again.
- If it's vertical, it won't meet the curve again.

However, it's important for elliptic curve theory that every line through two points on the curve gives you a third point: in the second case, the third point is taken to be whichever point the curve grazes past (it's a *double root*); in the third case, you say the third point is *the point at infinity*, which is called O.

This gives the curve some lovely algebraic (that's algebra in the abstract sense, not in terms of the "do the same thing to both sides" you do at school) qualities.

In particular, you can define a way to add points together, eventually making an abelian group. If a point A is on the curve, its reflection in the x-axis is $-A$. If you draw a line through two points (say, P and Q) and it crosses the curve again at a third point, R, "adding" the points P and Q together gives you $-R$.

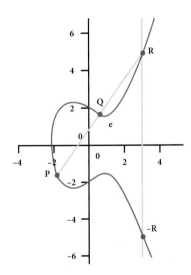

An elliptic curve demonstrates some fascinating algebraic qualities.

There are a few special cases involving the point at infinity, called O, because $P + (-P)$ never crosses the curve again, while $P + (-P) = O$. For similar reasons, $P + O = P$.

How about adding a point to itself? In that case, you draw the tangent to the curve at that point, and call R the point where that line crosses the curve again.

If you define addition this way, and call the point at infinity O, the points on the curve form an abelian group.

This also goes for just the points with rational coordinates. Which is all very nice, of course: woohoo! We made a group. Well done us, Fields Medal is in the mail. But why are elliptic curves important?

Well, first of all, they're important for their own sake. You can do fun math using them.

One of the Clay Millennium problems relates to a conjecture based on elliptic curves — if you can prove it, the Clay Mathematics Institute will give you $1,000,000, so feel free to have a crack at it!

There's also the modularity theorem, which was the key to Andrew Wiles's proof of Fermat's Last Theorem.

Their main uses, though, are in two related fields: in number theory, particularly for proving that a large number is prime, or for factorizing a number; and in cryptography, which exploits two facts about the group of points on an elliptic curve:

- It's very easy to add a point to itself over and over and over again, and

- It's very hard to work out how many times a point was added to itself in this fashion.

This is an example of a *one-way function*: something that's easy to do but very hard to undo by an eavesdropper — perfect for sharing secrets!

Cryptography is essential to the security of modern communications.

Like using a helicopter to go shopping, some math tools seem like overkill.

ELLIPTIC CURVE NOMOGRAM

Some things in math seem a bit like using a helicopter to go to the corner store — the tools used are far heavier-duty than the task required. On the other hand, the look on the face of the kid at the counter makes it worthwhile.

The elliptic curve nomogram is a perfect example. Using an elliptic curve and a ruler, you can multiply and divide numbers to whatever accuracy your printer can manage!

First, you write two scales on the curve, one in blue and another in red.

These scales are reciprocals of each other, meaning that, at any given point, the product of the red and blue values at that point is 1.

If you connect two numbers with the same color, you can read off their product in the other color where

PARKING ONLY

the line crosses the curve — you may need to adjust the decimal point.

If you connect two numbers with different colors, the line will cross the curve again at their quotient: reading off the blue number will give you the value of the fraction with the blue number on the bottom, whereas the red number will give the fraction with the red number on the bottom.

The curve can also be used to find square roots by clever use of tangent lines ... but I can't go into that right now, the helicopter's double-parked.

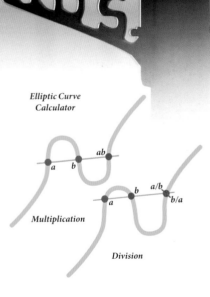

Elliptic Curve Calculator

Multiplication

Division

ELLIPTIC CURVE CRYPTOGRAPHY

The *Elliptic Curve Discrete Logarithm Problem* (or ECDLP, for short) sounds a lot easier than it is. Our hero, Alice, starts with a point, G, on an elliptic curve. She picks a secret big number n and works out $G + G + G + \ldots$, n times.

(That's nG if you're into the whole brevity thing.) Let's call that Q. Alice tells you G and Q. If you can work out n, you can read all of her secrets.

Luckily for Alice's secrets, finding n is extremely difficult, especially if the elliptic curves are defined on a *Galois Field* — simply using modulo arithmetic. Let's say Alice wants to send her friend

Alice and Bob pick a curve and a point, G

Bob picks a secret number, m

Alice picks a secret number, n

Bob works out mG (using elliptic curve arithmetic)

They tell each other their results

Alice works out nG (using elliptic curve arithmetic)

Bob works out $m(nG)$

Alice works out $n(mG)$

Because $m(nG) = n(mG)$ Alice and Bob have agreed on a secret point!

Alice needs to share certain information with Bob.

Bob a secret message. They can call each other up openly and say, "Let's use this elliptic curve, and this base point, G." Alice picks a secret big number n, and works out nG. She tells Bob (again, openly) the result.

Bob does the same thing with his own secret number, m, works out mG and tells Alice (still in the open!) his result. Alice and Bob have now established a common secret point — if Bob adds Alice's number to itself m times (his secret number), he gets mnG.

If Alice adds Bob's number to itself n times (her secret number), she also gets nmG — but nobody else has enough information to work out this answer! As long as Alice and Bob are sure nobody's been messing around with their numbers, they can safely use this as the basis of a code to communicate in complete secrecy.

I wonder what they're up to?

ANDREW WILES

Andrew Wiles (1953–) is one of the rare exceptions to the "no lone genius" rule
— and even he drew heavily on other people's work in finally achieving something
mathematicians had been attempting since 1637: he proved Fermat's Last Theorem.

Strictly, it was only at this point it became a theorem. Until then, it was a conjecture. He became fascinated by the "theorem" at age 10, coming across it in a library book, and loving that it

Andrew Wiles admires the Fermat Memorial in Beaumont-de-Lomagne, France.

was a statement so simple a schoolboy could understand it, but contained math so hard that nobody could prove it.

Wiles grew up in an academic environment. His father was Regius Professor of Divinity at Oxford University and prior to that had been Chaplain at Ridley Hall in Cambridge. It was in Cambridge that Andrew Wiles was born and, although Ridley Hall is a theological college and not part of Cambridge University, Wiles's early years were spent in the university town.

He attended The Leys School in Cambridge and went on to study mathematics at Cambridge University. He also studied at Oxford University and by the beginning of the 1980s was in America at the Institute for Advanced Study in New Jersey. He became a professor at Princeton University and

Fermat's image on a plaque marking his grave in Castres, France.

also spent some time in Paris and at Oxford before returning to Princeton, then settling at Oxford again in 2011.

Despite, or perhaps as a result of all his academic achievements, Wiles was never quite able to let go of that theorem he had first come across as a boy. As a grown-up mathematician, he set about proving it in a very sneaky way — in total secrecy (although he told his wife what he was up to).

He spent 6 years plugging away at the key missing link in the proof,

publishing old work in small batches so as not to arouse suspicion.

It might seem a scandalous omission that Wiles doesn't have a Fields Medal, but there's a good reason. The Fields Medal is restricted to mathematicians under 40. Wiles's proof was completed in 1994 — just a year too late.

The International Mathematical Union (IMU) gave him a plaque, though, which was nice of them. He was also given a knighthood for his work, becoming Sir Andrew John Wiles, KBE, FRS.

FERMAT'S LAST THEOREM REVISITED

I'm not going to pretend to understand Andrew Wiles's proof of Fermat's Last Theorem, but I can give you a flavor of what's behind it.

It's a proof by contradiction. It starts by assuming that Fermat's equation $a^p + b^p = c^p$ has a solution in the nonzero integers, with p a prime number at least 7. (The cases for 3, 5 and composite numbers have been proved as special cases).

Wiles at a conference in the Institute for Advanced Study, Princeton, in 2005.

If that assumption is true, then the elliptic curve $y^2 = x(x - a^p)(x + b^p)$ cannot be *modular* — a modular form being a type of complex analytical function. This much was known in the mid-1980s.

Wiles eventually proved the Taniyama-Shimura-Weil conjecture, now known as the modularity theorem (perhaps to make it easier to spell), that *every* elliptic curve, over the rational numbers, is modular — the missing link in the proof.

That's not just a case of "well done, Andrew." The conjecture was first stated in 1956 (by Taniyama) and, for a long time, was generally considered unprovable by experts in the field.

According to Simon Singh's excellent *Fermat's Last Theorem*, Wiles's supervisor, John Coates said it seemed "impossible to actually prove," while Ken Ribet, who worked extensively on the problem, counts himself as one of the many who considered it completely inaccessible.

The twist is this: although Fermat's Last Theorem has been proved,

John H. Coates was Andrew Wiles's supervisor.

it's certainly not Fermat's proof, assuming that he had one. The techniques Wiles used simply hadn't been dreamt up three and a half centuries before, when Fermat first claimed to have solved the problem. One thing that is for sure is that, given that Wiles's proof was 150 pages long, Fermat's margin was *certainly* too small to contain the marvelous proof.

Wiles announced his proof of Fermat's Last Theorem at the Isaac Newton Institute in Cambridge, England.

CHAPTER 15
THE TAMING OF THE MESSY

Gauss finds a missing dwarf planet behind the Sun, a London doctor saves Soho from cholera, and the Guinness company takes a zero-tolerance policy toward academic journals.

London and the River Thames were far from healthy in the middle of the 19th century.

THE MESSINESS OF DATA

Go and get some dice, ideally six, but one will do. Roll the six dice (or roll the one, six times) and keep track of which numbers come up. Did you get every number from one to six exactly once?

No ? That, in a nutshell, is the problem with statistics — nothing ever comes out exactly right (unless you manipulate the data, which is, of course, another kettle of fish altogether).

The question is, how do you go about making sense of data that — even when you have a good model — doesn't behave itself properly? You invoke the magic incantations of statistics.

Originally, statistics wasn't really a mathematical discipline, but a data-entry one — it was the collection and recording of information about the state. The word is often still used in that context, but in its math sense, it means the analysis and representation of data, which is what this chapter is all about: some of the effects statistical reasoning and presentation have had on the history of the world, and beyond.

Simply rolling dice proves that statistics seldom work the way you want them to.

Carl Friedrich Gauss correctly predicted the location of the dwarf planet Ceres.

Carl Friedrich Gauss (1777–1855) wasn't the first to look at how to minimize errors by combining multiple observations (work on that had gone on all through the 18th century), but he was one of the first to make spectacular use of it. In 1801, he correctly predicted the location of the dwarf planet Ceres, based on somewhat erratic observations made by the Italian monk Piazzi in 1801. For 40 days, the monk tracked the asteroid, only to lose it as it was eclipsed by the Sun. Gauss, alone, correctly predicted where and when it would emerge on the other side.

German 10-mark bank note featuring Carl Friedrich Gauss and his distribution model.

Gauss had noticed that, in many data sets, the frequency of an error of a given size was roughly proportional to the exponential of the square of its size. As a result, he figured out that the best way to make a prediction was to find the function that minimized the sum of the squares of the differences from it.

Unlike Laplace and Legendre, he managed to find the link between the two, develop the *normal distribution* — for many applications, the default way to model data — and demonstrate that, under certain conditions, this model was indeed the best available.

For that reason, it's called Gaussian distribution, and featured alongside Gauss on the German 10-mark note.

THE BROAD STREET PUMP

London had a bad year with cholera in 1854. It was a terrifying disease: the best explanation at the time of how it spread was via "miasma," the bad air circulating around the filthy and overcrowded city. Certainly, there was a lot of bad air and there were a lot of outbreaks of the disease.

On August 31, a particularly nasty one hit the overcrowded Soho district. Within 3 days, 127 people had died, and three-quarters of the population had moved away within a week.

This cartoon by Robert Seymour shows the cholera epidemic spreading in the form of poisonous air.

Enter physician John Snow, who was sceptical of the miasma theory. He didn't know the cause — this was nearly a decade before Louis Pasteur came up with the idea of "germs." Snow and the Reverend Henry Whitehead interviewed residents of the area to find out who had been affected, and plotted the deaths on a map of London.

British physician
Dr John Snow
(1813–1858)

Voronoi polygons overlaid on points plotted on John Snow's map of London's Soho district - statistics and infographics at work in the 19th century.

He also plotted the locations of the pumps used to provide the residents of Soho with drinking water, and invented what is now called the *Voronoi diagram.* He worked out which houses were closest to which pump and drew the boundaries. That was all the evidence he needed to trace the source of the disease.

Of the deaths Snow and Whitehead had identified, 61 lived closest to the Broad Street pump and were known to use it. Of the 10 closer to different pumps, five of the families interviewed said they generally used the Broad Street

pump because the water tasted better. Three of the other deaths were children who went to school in Broad Street.

The monastery right next door to the pump, though, was curiously unaffected. Divine intervention? Well, maybe. More likely, it was because they only drank the beer they brewed themselves, which didn't involve the pump at all.

On September 7, Snow spoke to the Board of Guardians of St James's Parish, who removed the handle from the pump the next day. The outbreak, already in decline, dwindled rapidly. By the end, the death toll stood at 616.

John Snow used statistics and infographics in Victorian London, a century before these kinds of visual tools became commonplace, to identify the cause of the epidemic and show the people in charge how to stop it.

Sadly, the Board of Guardians reattached the handle as soon as the outbreak had receded. Snow's theory that there was something in the water that caused cholera was simply too unpleasant to contemplate.

FLORENCE NIGHTINGALE

Florence Nightingale is known as a nurse but was also a pioneer of infographics.

As humanitarian figures go, Florence Nightingale (1820–1910) is one of the leading lights, at least of her age. She's best known as "The Lady With The Lamp" who organized a nursing contingent to help soldiers recover from wounds in the Crimean War. So why is she in a math book?

Convincing the British military to change *anything* is an uphill struggle. Even with the calamitous loss of life at the 1854 Battle of Balaclava, it was inconceivable that anything could be done differently — until Nightingale started assembling reports of the dead and wounded, presenting them in a coxcomb diagram, her own variant of the pie chart — it's sometimes called a Nightingale rose diagram.

With one graph, she managed to show the public, the politicians and the military leaders that many times more soldiers were dying in hospital from preventable diseases than from wounds received in battle.

Now, you're used to seeing infographics all over the place, but in Victorian England, they were relatively unknown. Certainly, getting MPs to read a detailed statistical report of the casualties would have been nigh-on impossible; getting them to glance at a graph and see the shocking extent of typhus, dysentery and cholera that were causing calamitous losses to the army was enough to persuade them into action.

That was vital — in her first winter in Crimea, the death rate at the field hospital was well over 40 percent. After the improvements in sewerage, sanitation and treatment she agitated for, that dropped to just 2 percent. A similar campaign into sanitation in rural India reduced mortality among soldiers from around 7 percent to under 2 percent.

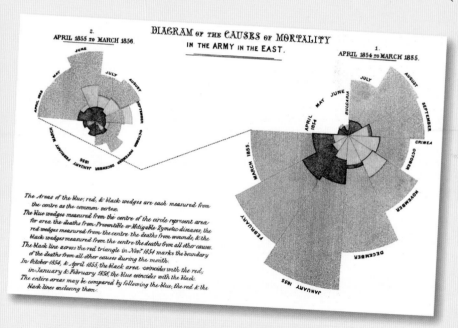

Diagram of the causes of mortality in the army in the East by Florence Nightingale.

Victorian London was
overcrowded, polluted
and had poor sanitation.

Nightingale also lobbied for effective sanitation in private houses in Britain, which some historians credit for increasing the average life expectancy by nearly two decades between 1871 and the mid-1930s.

Aside from being a pioneer of the visual display of information and meticulous record-keeping, Nightingale did probably more than anyone else to professionalize the career of nursing. She was the first woman to be awarded the Order of Merit, the first female member of the Royal Statistical Society, and an honorary member of the American Statistical Society. There are museums in her honor in London, Istanbul and at Claydon House in Buckinghamshire in England.

Statue of Florence Nightingale in Waterloo Place, Westminster, London.

THE GUINNESS COMPANY'S TRADE SECRETS

When William Sealy Gosset graduated from New College, Oxford in 1899, he got the job most graduates dream of — he went to work for Arthur Guinness & Sons in Dublin. As a statistician, he was concerned with checking the quality of the barley used in the brewing process.

British statistician William Sealy Gosset (1876–1937) worked on the problem of using small samples.

There was one big problem with this: most of the statistical techniques of the day involved huge numbers of observations — it was dubious, at best, to apply them to the handful of samples Gosset typically had access to. When Gosset spent a few terms on secondment with Karl Pearson, one of the fathers of statistics, they worked together on the small-samples problem. Pearson, although extremely helpful, didn't really see the point (he was a biologist; for him, there was no problem that couldn't be solved by taking a bigger sample.)

Gosset's other big problem was with his employers. A previous researcher at Guinness had published a paper and inadvertently revealed some of the company's trade secrets. The company didn't like that. The company didn't really even want its rivals to know it had a statistics branch.

As a result, it banned its employees from publishing any kind of paper, no matter what the contents, and no matter how important the underlying ideas.

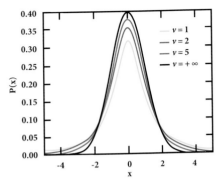

A graph showing the density function for several members of the Student "t" family.

Gosset's work involved monitoring the quality of barley used by the Guinness brewery.

Gosset's ideas were extremely important. In many fields, taking samples is very expensive, so making the best of little information is paramount. He pleaded with the brewery, explaining that his work really didn't contain anything secret. It didn't even need to be about brewing. "Okay," said the board, "but you can't use your own name, or everyone will be begging us to let them publish."

"Fine," said Gosset, and published under the pseudonym Student. His best-known work is called Student's t-test, rather than Gosset's. He wasn't bothered about the lack of credit — he claimed that Fisher — the other father of statistics — would have discovered it all in any event.

> "Just because nobody complains, doesn't mean all parachutes are perfect."

> Benny Hill

ABRAHAM WALD AND THE LOST PLANES

The USAF was losing an awful lot of valuable planes and even more vital pilots. It was the middle of World War II and losses over occupied Europe were mounting alarmingly.

They looked at the planes that had limped back to England from sorties. The thinking was, by patching up the areas that came back in tatters, they could save more planes.

Abraham Wald, the brilliant Hungarian mathematician, one of the lucky ones who'd escaped before the

Abraham Wald (1902–1950) fled his native Hungary to escape the Nazis and worked with the Allies to improve the survival rate of aircrew.

During World War II, aircrew often returned from missions in heavily damaged aircraft. Statistical analysis showed how to improve their chances of survival.

Nazi invasion, shook his head sadly. I like to think of him calling the Colonel a buffoon: "Those, sir, are the worst possible areas to put heavy armor."

Wald had spotted two things: first, that no weaponry at the time could accurately hit any part of a plane, so a fighter was about as likely to be hit in any one place as any other.

Second, if a pilot could make it home with a damaged fuselage, for instance, it followed that the fuselage was a bit of a luxury, as far as flying was concerned and it didn't need extra armor because it wasn't actually critical to the survival of the aircraft.

The correct place to put the extra armor was in the places where *none* of the surviving planes had been shot. Since every part of the plane was just as likely to be hit, the ones that *didn't* make it home were likely to have suffered critical damage there.

Not content with saving the lives of hundreds of airmen by helping the top brass understand survivorship bias, Wald went on to model how much damage a plane could suffer in each section, and how likely it was to incur it on any given run — meaning the commanders could plan their sorties so as to minimize their likely losses.

BUFFON'S NEEDLE

The year is 1730-something and Georges-Louis Leclerc,
Comte de Buffon, is in his study with a big box of needles.
He picks one up, carefully spins it into the air, and
— when it lands — he records whether it landed
overlapping the edge of one of his floorboards.

Being a French aristocrat, Buffon has an excuse
for pursuing eccentric hobbies. One of his
hobbies considered eccentric is his questioning of
the processes by which different kinds of
animals came about — although he believes
in the story of Adam and Eve, he's one of
the first to begin making the kinds of
enquiries that will, one day,
lead Darwin to his discoveries.
There is little distinction
between the branches
of science in the 18th
century, so it's natural
for Buffon to be
interested in math
and probability,

Portrait of Georges-Louis
Leclerc, Comte de Buffon,
1753, Musée Buffon,
Montbard, France.

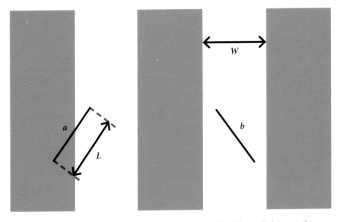

Needle a has landed across the edges of two floorboards, while needle b is entirely
clear of the edges. Counting the proportion of needle that land
in each of the configurations can give an estimate for π.

too — which is why he's throwing the contents of a sewing kit on the floor. He has a theory, one he's worked out very carefully: if you throw enough needles at the floor, making sure they land in random directions, the proportion of needles (each of length L) landing across the edges of the floorboards (which are W wide), ought to be:

$$2L/\pi W$$

... meaning, with a little bit of algebra, you can get a good estimate for π this way.

The Comte de Buffon has just invented the field of *geometric probability* and, with the same idea, come up with the first *Monte Carlo experiment*.

It's not recorded whether Buffon ever performed the experiment — aside from his interests in math and biology, he was also the King's gardener, so he was a busy man — but one man who certainly did was the Italian mathematician Lazzarini, who tried exactly this in 1901, and got an astonishingly good result.

Number of repetitions R	Number of crossings C
100	53
200	107
1,000	524
2,000	1,060
3,000	1,591
3,408	1,808
4,000	2,122

Theoretical values for Buffon's needle experiment, which Buffon may never have attempted himself.

Lazzarini tossed 3,408 needles, each five-sixths of the width of his floorboards — so the probability of each of his needles lying across the line would be: **5/3π**

Of the 3,408 needles, 1,808 landed on the lines, giving him an estimate of π that was 355/113 = 3.1415929.

The decimal value of π to seven decimal places is 3.1415927, so that's not a bad guess for a few hours' work. You might almost consider it too good to be true.

I'm not accusing Lazzarini of manipulating his results. But it's worth stopping to ask, why did he pick 3,408

needles to throw? Why not 3,000, or 4,000, or even 3,400? The reason is, Lazzarini knew what fraction he was aiming for — 355/113 is a well-known, excellent approximation for π — there isn't a better one involving numbers under 10,000.

Lazzarini knew that, if he threw 213 needles in this experiment and 113 landed as he wanted, he'd get exactly that fraction. The chances of that happening are about one in 20. However, if he didn't get that, he could carry on and throw another 213 needles and see if he ended up with a double 226. If you keep going like this, the probability of you *not* reaching the target fraction at some point drops slightly with each batch. Lazzarini ultimately threw 16 bundles of 213 needles, making 3,408.

He got a bit lucky (the chance of hitting the jackpot on or before your 16th batch is about one in four), but all the same: when you're doing an experiment, you probably shouldn't start with a target in mind, or you'll affect your results.

In nature, pi crops up regularly. The winding of a river, for example, is referred to as its sinuosity — its length divided by the direct distance from its source to its mouth. Under ideal conditions, the sinuosity count is about 3.14.

GALTON'S OX

Sir Francis Galton (1822–1911) was a very clever man with some extremely obnoxious ideas. He was a great proponent of eugenics (in fact, he came up with the word), which rather clouds his achievements. It's a pity; even people with awful ideas can still come up with good ones.

For example, in studying how human ability might be inherited, Galton came up with experimental ideas still used today. Studies of identical twins, for example, are the mainstay of behavior genetics, since if two people have identical sets of genes, any differences between them can't be genetic ones. He was also a pioneer of statistical techniques in heredity, using correlation and regression to model traits. He invented the standard deviation, a measure of how far a measurement can be expected to stray from the mean. He invented the "quincunx," a sort of pinball machine you can use to show what the normal distribution looks like. He popularized the idea of "regression to the mean," the idea that exceptional performances (either good or bad) tend

Sir Francis Galton did some extremely valuable work pioneering statistical techniques, but also had some decidedly odd ideas in other areas.

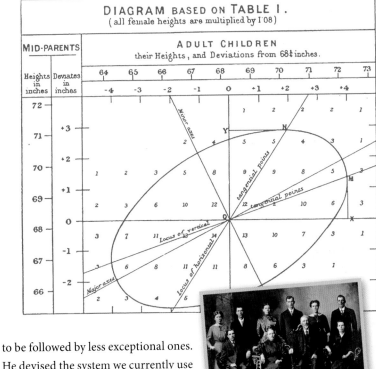

Galton's correlation chart formed part of his studies of genetic traits within families.

to be followed by less exceptional ones. He devised the system we currently use for classifying fingerprints.

He's probably most famous, though, for a visit he made to a fair. If you've ever been to a church fete, the chances are you've seen a "guess how many candies are in the jar" competition — whoever is closest to the correct number wins the jar. At this particular fair, it wasn't a "guess how many candies are in the jar" contest, but "guess the weight of the ox."

*Rather than trying to
win an ox at the fair,
Galton wanted to study
the guesses people made
about its weight.*

Galton didn't want an ox. What would he want with an ox? He wanted the guesses. Out of almost 800 people who paid sixpence each for the chance to take part, nobody got the right answer. However, he found that the median of the guesses (1,207 pounds, or 547 kg), which he thought captured the *vox populi*, was within 5 kilograms of the correct answer, which was 1,198 pounds, or 543 kg. Even better, the mean of the guesses turned out to be 1,197 pounds.

THE CURSE OF SPORTS ILLUSTRATED

In American sporting folklore, there's a superstition that says: soon after you appear on the cover of *Sports Illustrated* magazine, your performances will take a nosedive. The data appears to back this up. Charting the relevant statistics shows that, generally speaking, a player's form does tend to drop after they've been featured in the magazine.

What is this strange magic the editors of a sporting magazine have over players of baseball, basketball, football, tennis, ice hockey and the rest? How can they harness their power for good instead of evil?

Sadly — or perhaps fortunately — there's no such power. The drop in form can be explained simply by regression to the mean. If you've been performing well enough to be featured on the cover of the magazine, the chances are you've been performing unusually well. Sometimes, that's because your underlying talent has suddenly improved. In many more cases, you've just been on a lucky run — and it inevitably comes to an end.

This is why stereotypical bosses think that yelling at people to make them improve works, whereas praising them for good work is counterproductive. After either of those interventions, the employee who has been under- or over-performing compared to their usual effort tends to revert back to their usual performance. The manager wrongly attributes this to the intervention rather than basic statistics. Managers, eh?

CHAPTER 16
MODERN HEROES

Some of the heroes of the 20th and 21st centuries get a shout-out.

PAUL ERDÖS

He'd wander into the department, disheveled after a flight, a suitcase of meager belongings in one hand, a suitcase of papers in the other, and announce "My brain is open."

Erdős (1913–1996) would collaborate with anyone who would talk to him, and in this manner, wrote or co-authored somewhere in excess of 1,500 papers in his lifetime.

If you've ever played the game "Six degrees of Kevin Bacon," you'll know that it's possible (in fact, usually fairly easy) to link any active film actor to Kevin Bacon by way of people with acting credits on the same movie in six links or fewer — Bacon, famously, has worked with almost everyone in Hollywood. In math, the same idea applies to Erdős. Many mathematicians proudly boast of their "Erdős Number," how many co-authorships it takes to get from them to Erdős. (Mine, I believe, is five, although it may be lower. I co-authored with Eric Priest; Eric

Paul Erdős at a student seminar in Budapest in 1992.

wrote several papers with Mitch Berger; Mitch and Keith Moffatt collaborated, while Moffatt worked with George Lorentz — who wrote "On the probability that n and $g(n)$ are relatively prime" with Erdős in 1959.) There are 511 known people with an Erdős number of 1.

His big mathematical trick was the "probabilistic method," which works by examining a random object in a collection and showing that the probability of it having a particular property isn't zero. If that's the case, there must be an object in the collection with the property — which means you can prove that there's an object with the property without specifying which one!

Mathematician Ronald Graham (center), has an Erdős number of 1.

Apart from his obsession with doing math — he never stopped traveling from conference to department to friendly household, drinking vast quantities of coffee as he went — Erdős was renowned for setting challenges just beyond the current threshold of mathematical knowledge, and offering prizes for their solutions.

Many of these are still open, including the Collatz conjecture — if you can solve that, get in touch with Ron Graham and claim your $500 prize!

THE COLLATZ CONJECTURE

Pick a number, any number. If it's even, halve it; if it's odd, multiply it by 3 and add 1. Repeat the process with the result until you get into a cycle.

For instance, if you start with 18, you would halve it (to get 9), then treble it and add 1 (to get 28). Halve it: 14. Halve it: 7. Treble and add 1: 22.

Then:

11, 34, 17, 52, 26, 13, 40, 20, 10, 5, 16, 8, 4, 2, 1, 4, 2, 1 …

… there's a cycle!

The Collatz conjecture is that whatever number you start from, you eventually reach 1, and has remained resolutely unproven since 1937. Computers have checked every number up to the quadrillions and found them to work, but that's not the same thing as a proof, not at all. To disprove it, you just need to find one counter-example: a number that either keeps on going up, or that gets stuck in a different cycle. To prove it, you need to show that both of those things are impossible.

If you're bored one afternoon, try starting from 27. Bring plenty of paper.

German mathematician Lothar Collatz (1910–1990) posed his conjecture in 1937.

The Indian mathematician Srinivasa Ramanujan (center) together with his colleague Godfrey Harold Hardy (extreme right).

SRINIVASA RAMANUJAN

Getting crank letters goes with the territory of being a mathematician. Most math departments have a wonderfully efficient secretarial staff who weed out the worst of the scary people.

Some claim to have proved Fermat's Last Theorem by taking logarithms, whereas others have discovered the hidden mystical significance of the duodecimal expansion of π.

Luckily, in 1913, the University of Cambridge had no such system in place, and a letter from India landed on G. H. Hardy's desk. Hardy himself was a pure mathematician of some repute, especially in analysis and number theory, and — despite his general hope that his mathematics should never be sullied with anything as base as an *application* — is known for the Hardy-Weinberg principle, a result involving stable populations in biology.

In any case, the letter that came to Hardy was nine pages of densely packed equations and identities, some of which Hardy found familiar, and others that seemed incredible. Hardy suspected a clever and intricate fraud.

There was no fraud. The letter came from Srinivasa Ramanujan (1887–1920), a self-taught mathematician from Madras (now Chennai, India). Hardy eventually decided that some of Ramanujan's results on continued fractions had to be true, reasoning that they were too bizarre to have been invented.

The following year, Ramanujan came to England to work with Hardy. Plagued by ill-health, he managed 6 years of collaboration before falling to what was presumed to be tuberculosis, although it may have been liver disease.

Ramanujan was a complete contrast to Hardy. His insights seemed intuitive, coming from nowhere, and generally without proof — completely at odds with Hardy's insistence on rigor. Yet the pair made a great team; once the insight had been established, a proof could eventually be found.

Among other things, Ramanujan discovered identities involving the hyperbolic secant, mock theta functions and infinite series for π, one of which gives the remarkable approximation of:

$$\frac{9801\sqrt{2}}{4412} = 3.14159273$$

correct to about one part in 40,000,000.

To nonexperts, he's best known for an off-hand comment to Hardy, who had come to visit him in hospital. Hardy remarked that he'd taken cab 1729, which wasn't a very interesting number. Ramanujan immediately pointed out that it's the smallest number that can be written as the sum of two cubes in two different ways ($10^3 + 9^3 = 12^3 + 1^3$) — leading to the idea of a taxicab number.

Incidentally, it's not known whether any number can be written as the sum of two fifth-powers in two different ways — that's an open problem you might like to tackle on your next coffee break!

Bust of Srinivasa Ramanujan, Indian mathematical savant in the garden of Birla Industrial and Technological Museum.

GRIGORI PERELMAN

I hesitated about including Grigori Perelman (1966–) in this list of great recent mathematicians — there's no question that he's an amazing mathematician, but he's also extremely reclusive and antipathetic towards recognition. Grigori, if you're reading this, I apologize.

If Perelman was the sort to claim fame, his claim to fame would be "I'm the only person to have solved one of the Clay Millennium problems." Like Hilbert's program in 1900, at the turn of the 21st century, the Clay Institute suggested that there were seven unsolved problems in mathematics worthy of a million-dollar prize:

Grigori Perelman at Berkeley in 1993.

- P vs NP, a question about the complexity of algorithms;
- The Hodge Conjecture, which concerns projective algebraic varieties;
- The Riemann Hypothesis, which concerns an infinite sum in complex analysis — and was actually Hilbert's eighth problem — with deep consequences for number theory;
- The Yang-Mills existence, a problem in quantum theory;
- The existence of smooth solutions to the Navier-Stokes equations, the bane of my fluid dynamics studies;
- The Birch and Swinnerton-Dyer conjecture, which concerns rational solutions on elliptic curves; and
- The Poincaré conjecture, which is about the topological character of hyperspheres.

Of the seven, only two have solutions that have been even considered by the Clay Institute: Cho, Cho and Yoon's proposed solutions to the Yang-Mills problem, according to the Institute's commentary, "simply do not go far enough." Grigori Perelman's 2003 proof of the Poincaré conjecture, on the other hand, was awarded the prize in 2010.

Not that Perelman accepted it: he felt that the work of Richard Hamilton toward the proof was at least as deserving of the credit, and said that the award was unfair. He also declined to receive the Fields Medal in 2006, the only person to turn down the honor.

Turning down a Fields Medal is hard for a mathematician to understand. Turning down a million dollars you've been awarded is hard for almost anyone to understand. But Perelman has his reasons. For him, the proof is enough. He doesn't want to be permanently on display "like an animal in the zoo."

Irritated at the fuss surrounding his awards, he has completely withdrawn from the mathematical community — although there are rumors he's working on other things (perhaps Navier-Stokes), while living with his mother in St Petersburg, ignoring journalists and anyone else who bothers him.

To me: that's a true mathematical hero. For him, the proof is its own reward. Good luck to him.

Perelman lives quietly in St Petersburg, Russia.

I'm not a hero of mathematics. I'm not even that successful; that is why I don't want to have everybody looking at me.

Grigori Perelman

EMMY NOETHER

Emmy Noether (1882–1935) was one of the most influential mathematicians of the 20th century.

As an algebraist, she revolutionized the study of (mathematical) rings, fields and algebras — all three have technical meanings rather different from the common English usage, and underpin a huge swathe of pure mathematics. Meanwhile, as a theoretical physicist, she connected the ideas of symmetry and conservation. The connection is now known as Noether's theorem — for once, a theorem named after its discoverer! The idea is that any symmetry you can observe in a physical system corresponds to a conservation law. For example, if you drop a cannonball from a tall tower, it doesn't matter whether you do it today or tomorrow — it is invariant under time, which leads to

Noether's theorem is of huge importance to physics.

the conservation of energy. It doesn't matter whether you do it from this tower or the one next door — it's invariant under space, which leads to the conservation of momentum. It's an extremely neat (if tricky) idea.

Noether's main maxim was abstraction: the more you can isolate techniques, operations and so on from their applications, the more useful they become — because you can then apply them to apparently unrelated fields in unexpected ways (she was, eventually, dismissive of her own thesis, calling it a "jungle of equations" — presumably a concrete jungle).

Noether worked for many years with David Hilbert and Felix Klein at Göttingen — although for the first 4, she was barred from

Noether studied at the University of Erlangen in Bavaria. She dismissed her thesis as "a jungle of equations."

becoming a member of the faculty for being a woman (in the meantime, she took over some of Hilbert's lectures). As the faculty's sexism delayed the start of her career at Göttingen, the government of the day's anti-Semitism hastened its end, and she left for America in 1933. She died two years later after suffering an ovarian cyst.

I can testify that [Noether] is a great mathematician, but that she is a woman, that I cannot swear.

E. Landau

MARYAM MIRZAKHANI

Maryam Mirzakhani is presented with the 2104 Fields Medal in Seoul by South Korean President Park Guen-hye. Mirzakhani was the first woman to receive the award.

I thought, for a split second, I was going to have to write about someone younger than me, which would have been galling; happily, Maryam Mirzakhani (1977–) is about 6 months my senior, which explains why she has achieved so much.

Iranian by birth, Mirzakhani lives and works in the United States. A professor at Stanford, she was one of the four Fields Medal laureates in 2014, winning for her work on understanding the symmetry of closed surfaces in moduli spaces.

What's a moduli space? I'm so glad you asked. It's a space where the points aren't your typical x, y, z coordinates, but instead algebraic or geometric objects of some sort — which can then somehow be given coordinates.

For example, if your objects were ellipses, you would be able to classify them by the lengths of their axes (in this case, you don't care where they are or how they're angled), so you would need two parameters to determine any ellipse — these would be the coordinates of a moduli space.

Two ellipses with similar measurements would have coordinates close to each other.

Mirzakhani doesn't work (as far as I know) on the moduli spaces of ellipses. She works on the moduli spaces of Riemann surfaces, and showed that the closures of complex geodesics (particular kinds of curves and surfaces) in moduli spaces are "surprisingly regular" — that is, they're not fractal or otherwise irregular — echoing work in less complicated spaces from the 1990s by Marina Ratner.

Mirzakhani is a professor of mathematics at Stanford University.

NICOLAS BOURBAKI

Uniquely in this heroes section, Nicolas Bourbaki doesn't have any dates attached to him. I suppose (c. 1934–) would just about be true, but it's a dangerous thing, assigning dates to someone who doesn't really exist.

The name "Bourbaki" was taken from the 19th-century French general Charles-Denis Bourbaki.

Instead, Bourbaki is a secretive group of mathematicians who first met in the mid-1930s with the goal of creating a proper math book. Not quite the "prove everything from the ground up" sort of thing that Frege and Russell and Whitehead had tried; it has no pretensions to being complete, just abstract and as general as possible. And, of course, to proceed rigorously from axioms.

Although it's a secretive group, some of its past and current members are known: Jean Coulomb, Jean Dieudonné (often the group's spokesman) and André Weil were founder members, while other Bourbakians included Alexander Grothendieck and Cédric Villani (both Fields Medallists).

So far, the group has produced nine books, each with the kind of title the average mathematician would look at and say "Oh! I know what that is!" before looking at the content and saying "… or rather, I thought I did."

They were writing *that* kind of math book.

Bourbaki did introduce several important words and symbols. It's hard to imagine functional analysis without the words injective, surjective and bijective, and the symbol for the null set, Ø, is another Bourbaki-ism. Lastly, there's the marginal symbol of "dangerous bend," inspired by road signs — if you see this in a Bourbaki book, it means that what you're about to read is harder or subtler than you might be expecting.

The group has an office at the École Normale Supérieure in Paris.

The Bourbaki approach is to take everything in its logical order and build results up in a coherent and undeniable way — everything follows from what has gone before (this was, in part, a reaction against the ideas of Poincaré, who was much more of a "let it all flow freely" sort of chap). The group would meet several times a year and loudly argue before voting on each line of the text. There was nothing but text, of course.

Meetings at the office are held in secret several times a year.

JOHN HORTON CONWAY

I know they're just artefacts of a system of rules — if a square has two neighbors, it stays as it is; if it has three neighbors, it will become black (live) in the next iteration; otherwise it becomes white (dies) — but they look for all the world like ants marching slowly, diagonally across the screen.

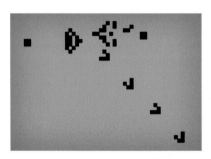

Gosper's Glider Gun is one of the many fascinating patterns generated by the Game of Life.

The Gosper's Glider Gun is just one of the patterns that comes out of the very simple rules of John Horton Conway's *Game of Life*, one of the very first things anyone learning to code should implement.

It's not the only game Conway (1937–) invented, and certainly not the most strategic (*Game of Life* is a zero-player game, in the sense that once the initial board is set up, nobody has to make any decisions). He also invented *Philosopher's Football* (two players draw lines on a grid to try to move the ball into their opponent's goal) and *Sprouts*, an infuriating game of doodles.

Aside from inventing games, Conway analyzed plenty, including the Soma cube and peg solitaire, and wrote several books on the subject. Oh, and he came up with a completely new number system (the surreal numbers), and a way of writing extremely large numbers (chained arrow notation) and the Doomsday algorithm for working out the day of the week for any given date.

Let's not talk about his distinguished career in serious math, because who cares about serious math? Well, if you insist. He proved Waring's conjecture

*John H. Conway
has popularized
mathematical games
but also makes serious
contributions to
number theory.*

(every integer can be written as the sum of 37 fifth powers), worked on knot theory and group theory, and showed, astonishingly, that if quantum experimenters can decide freely what they want to measure, the elementary particles being measured must be able to choose their properties just as freely — meaning, "if experimenters have free will, so do elementary particles."

Originally from Liverpool, England, he is now a professor at Princeton.

MARTIN GARDNER

My final modern hero of recent mathematics, possibly fittingly, isn't really a mathematician: instead, it's someone who inspired countless mathematicians — as well as magicians and chess players — by writing engagingly about the subject.

Mathematics writer Martin Gardner had little formal math education.

In 1956, after writing several articles about paper-folding for *Humpty Dumpty* magazine, Martin Gardner (1914–2010) submitted one on an especially nice structure, the hexaflexagon, to *Scientific American*. As the name suggests, the hexaflexagon is a hexagon that you fold and unfold to reveal unusual and interesting patterns.

Simple enough for a child to understand the idea (I remember it showing up in books of my childhood) but complicated enough that the math behind it isn't obvious (it's related to the Möbius strip), the article on it was an enormous and immediate hit.

The editor asked Gardner to produce more of the same: his *Mathematical Games* column ran for the next 25 years.

In it, he covered many of the topics I've mentioned in this book — fractals, Conway's *Game of Life*, public-key cryptography

The first cover of Humpty Dumpty *in 1952.*

— and hundreds of others besides. His columns have been compiled into many books (if you're looking for something to read after this, I strongly recommend them), but his best-selling effort was his 1960 book, *The Annotated Alice*, explaining many of the riddles and much of the word-play in Lewis Carroll's work.

In 1993, a conference in his honor

was held in Atlanta, Georgia, and repeated in 1996. Since then, the Gathering for Gardner, a get-together of people who share his interests in math, puzzles and magic — or anything else he might have written about — has taken place every 2 years.

For me, the most remarkable thing about Gardner is that he was never a card-carrying mathematician — calculus flummoxed him in high-school and he took no more classes after that. Which goes to show there's an enormous difference between the math you have to learn in school and the math that's genuinely fascinating to everyone.

A hexaflexagon, shown with the same face in two configurations.

You're in the same racket as I am — you just read books by the professors and rewrite them!

Isaac Asimov
on Martin Gardner

INDEX

8714

ACKNOWLEDGMENTS

The publisher would would like to thank the following for their kind permission to reproduce their photographs:

The images on the following pages are Public Domain:

p7, p8, p15, p23, p32, p33, p34, p39, p41, p55, p58, p61, p68, p70, p72, p75, p81, p93, p94, p95, p101, p136, p141, p145, p146, p147, p151, p154, p158, p17,8, p180, p181, p194, p204, p206, p209, p211, p211, p213 (left), p213 (top right, p215, p216, p217, p218, p220, p223, p224, p225, p236, p243, p249, p251, p255, p259, p260, p263, p271, p278, p284, p287, pp291, p294–295, p298, p302, p303, p306, p307, p341, p347, p349, p350, p352, p353, p354, p356, p358, p360, p364, p374, p378, p382, p387.

All other images are iStock.com unless stated otherwise:

Cover: Marina Sun/Shutterstock.com

p13 Ben2, p24 Almare, p25 Lakey, p40 (top) Shutterstock.com, p40 (bottom) Claus Ableiter, p46 Stockholms Universitetsbibliotek, p47 Board of Regents of the University of Oklahoma, p49 Dreamstime.com, p50 (bottom) Giorgio Gonnella, p59 Aleph, p71 Wellcome Trust, p73 Benh Lieu Song, p116 Hans A. Rosbach, p118 Andrew Dunn, p123 Andrew Dunn, p124 Hajotthu, p126 Chris 73, p131 MJCdetroit, p133 Japs 88, p135 Cormullion, p139 DXR, p148 wikispaces, p160 Ad Meskens, p160–161 (bottom) Arnold Reinhold, p161 (top) Roger McLassus, p171 JP, p182–183 Noah Slater, p190 Bjørn Smestad, p197 ArtMechanic, p205 German Federal Archive, p208 Getty Images, p210 Andrew Dunn, p213 (bottom right) Konrad Jacobs, p222 Allan J. Cronin, p230 Dave Fischer, p239 Godot13, p240 Wellcome Trust, p248 Gryffindor, p250 Stanisław Kosiedowski, p252 Stako, p258 (top right) George M. Bergman, p270 Ibigelow, p279 Autopilot, p296 Sailko, p308 Lmno, p312 (top) Raul654, p314 British Ministry of Defence, p318 German Federal Archive, p319 David Monniaux, p325 Wolfgang Beyer, p327 NOAA, p331 Predrag Cvitanović, p340 Klaus Barner, p342 C. J. Mozzochi, Princeton N.J, p342 Renate Schmid – Mathematisches Forschungsinstitut Oberwolfach, p348 Deutsche Bundesbank, p367 (top left) Steve Lipofsky www.Basketballphoto.com, p367 (bottom left) SD Dirk, p370 Kmhkmh, p371 Che Graham, p373 Konrad Jacobs, p376 George M. Bergman, p379 Akriesch, p380 Lee Young Ho/Sipa USA, p383 (top left) Encolpe, p383 (bottom right) Marie-Lan Nguyen, p384 LucasVB, p385 Thane Plambeck, p386 Konrad Jacobs.